# When the River Rises
## Flood Control on the Boise River
## 1943–1985

# *When The River Rises*
## Flood Control on the Boise River 1943-1985

Susan M. Stacy

Program on Environment and Behavior
Special Publication No. 27

Institute of Behavioral Science
Natural Hazards Research and Applications Information Center
University of Colorado

Published jointly with
College of Social Sciences and Public Affairs
Boise State University
1993

Front cover: Corps of Engineers aerial photograph dated May 1950 courtesy of U.S. Army Corps of Engineers, Walla Walla District,. The proposed bridge was the 16th Street Bridge, now called the Americana Bridge, which was completed in 1951. Boise's "proposed channel improvement" above and below the new bridge was not implemented, although fill at the north bank between the bridge and the Union Pacific railroad bridge eventually narrowed the channel. Cover design: Pete Wilson, DesignWorks

University of Colorado
Natural Hazards Research Applications and Information Center
Program on Environment and Behavior
Institute of Behavioral Sciences
Boulder, Colorado, 80309–0482

Boise State University
College of Social Sciences and Public Affairs
Boise, Idaho 83725

©1993 by the Institute of Behavioral Sciences, University of Colorado
All rights reserved.
Pubished in 1993.
Printed in the United States of America.
Designed by Gary Richardson, RENAISSANCE communications.

*Library of Congress Cataloging-in-Publication Data*

Stacy, Susan M., 1943–
    When the river rises : flood control on the Boise River, 1943–1985/ Susan M. Stacy.
        p. cm. -- (Program on environment and behavior) (special publication ; no. 27)
    "Published jointly with [the] College of Social Sciences and Public Affairs, Boise State University."
    Includes bibliographical references (p.    ) and index.
    ISBN 1-877943-08-8 : $15.00
    1. Flood control--Idaho--Boise River--History. 2. Floods--Idaho--Boise River--History. 3. Water resources development--Idaho--Boise River--History. I. Boise State University. College of Social Sciences and Public Affairs. II. Title. III. Series. IV. Series: Special publication (University of Colorado, Boulder. Natural Hazards Research and Applications Information Center) ; no. 27.
TC425.B57S72 1993
363.3´4936´0979628--dc20                                   92–31277
                                                                            CIP

To Mom and Dad

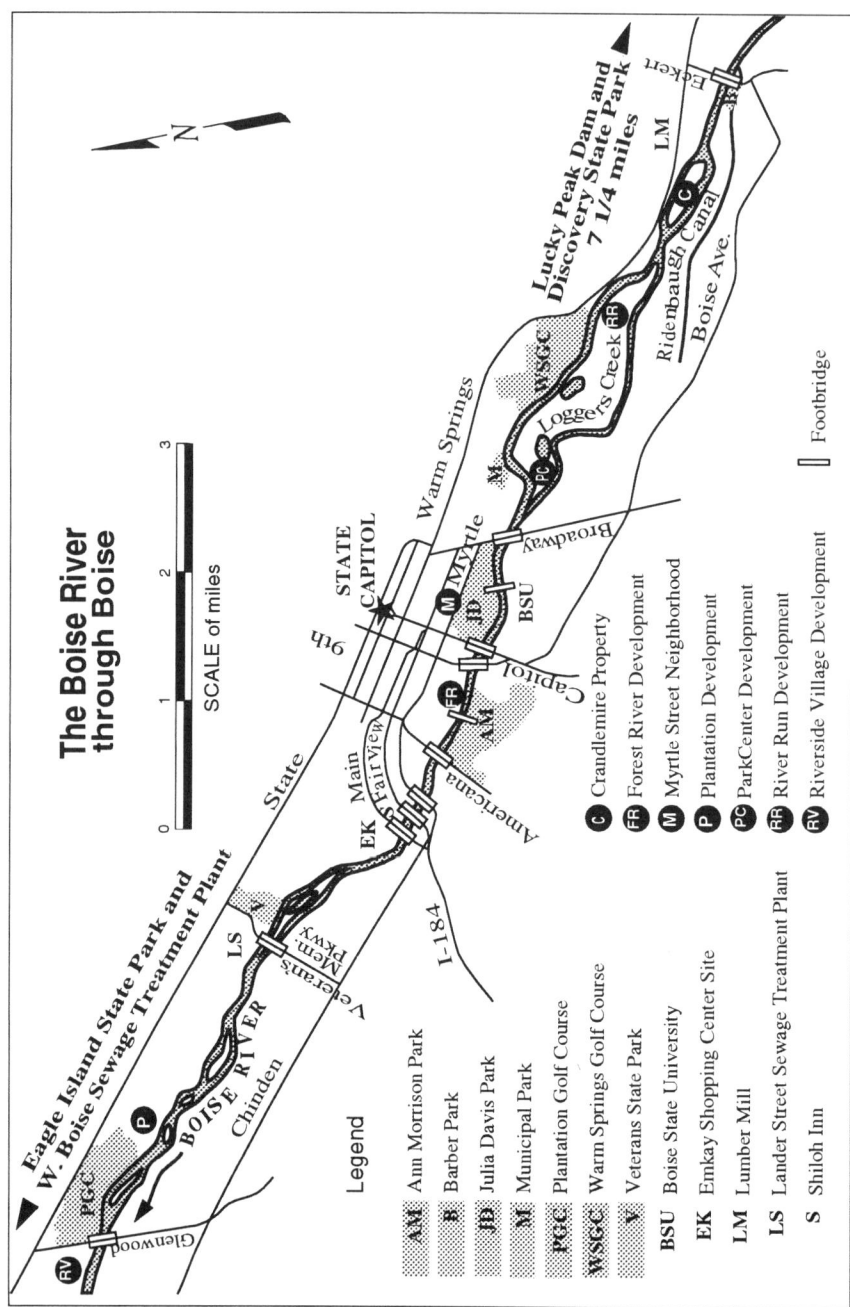

# Contents

| | |
|---|---|
| **Preface** | xiii |
| **Acknowledgements** | xv |
| **Before You Start** | xviii |
| **Introduction** | xxi |

**Chapter One: When Were the Floods?**   1
- The 1876 Flood: Early Settlement   2
- The 1896 Flood: Irrigation Expands   5
- The 1943 Flood: Federal Reclamation   8

**Chapter Two: After the Flood**   17
- Local Responses   17
- Federal Responses   22

**Chapter Three: Lobbying for Lucky Peak**   24
- Sweepy-Weepy and the Mountain Home Desert   24
- The Corps Plans Lucky Peak   29
- SWIWCP and The Corps Justify Lucky Peak   32
- Congress Approves Lucky Peak   35
- The Corps Builds Lucky Peak Dam   40

**Chapter Four: Conflict over Boise River Water**   44
- Constraints on the Corps   45
- Fish and Game Fight for Minimum Flow   51
- Valley Farmers Fight Flood Damage   55

| | |
|---|---|
| Boise Seeks Second Outlet Tunnel | 58 |
| Changing the Rules | 62 |

## Chapter Five: The Greenbelt     67
| | |
|---|---|
| Boise River Aesthetics, 1955-1965 | 68 |
| The Greenbelt Grows, 1965-1975 | 74 |

## Chapter Six: Developing the Flood Plain     80
| | |
|---|---|
| National Flood Insurance Program | 80 |
| ParkCenter | 83 |
| River Run | 87 |
| New Metro Plan Policies | 94 |
| The Boise River Plan | 96 |
| Amending the Plan | 99 |
| Economic Development and Garden City | 100 |
| The Bald Eagle Ideal | 103 |

## Chapter Seven: The Flood of 1983     106

## Chapter Eight: When the River Rises     119
| | |
|---|---|
| Channel Capacity | 119 |
| Diffusion of Management Responsibility | 121 |
| Ultimate Responsibility? | 122 |
| Urbanization of the Flood Plain | 125 |
| Valuing Public Goods | 125 |
| Western Water Use | 128 |

## Appendices     131
| | |
|---|---|
| Appendix A | 131 |
| Appendix B | 137 |

## Notes     139

## Bibliography     167

## Index     181

Contents      ix

# Illustrations

Figures and Maps

| | |
|---|---|
| Map of Boise River through City of Boise | vi |
| Boise River Watershed Map | xii |
| Strawberry Glen River Flows | xix |
| Aerial View of Boise River and Its Source | xx |

| | | |
|---|---|---|
| 1.1 | Plat of Boise in 1885 | 3 |
| 1.2 | Milton Kelly | 4 |
| 1.3 | Lumber mill | 6 |
| 1.4 | Broadway Bridge in 1896 | 7 |
| 1.5 | Diversion Dam in 1909 | 9 |
| 1.6 | Arrowrock Dam in 1916 | 9 |
| 1.7 | Plantation Golf Course in 1943 | 11 |
| 1.8 | William Welsh | 12 |
| 1.9 | Town of Eagle in 1943 | 13 |
| 1.10 | Linder Bridge in 1943 | 13 |
| 1.11 | Plantation Golf Course in 1943 | 14 |
| 1.12 | View of Boise River in 1943 | 16 |
| 2.1 | Flood view in 1936 | 18 |
| 2.2 | Flood scene between Boise and Eagle | 19 |
| 2.3 | Austin Walker | 20 |
| 2.4 | Notus Bridge in 1943 | 21 |
| 3.1 | Harry W. Morrison | 24 |
| 3.2 | "Valleys of Tomorrow" | 25 |
| 3.3 | Map of Mountain Home Project | 28 |
| 3.4 | Lucky Peak Dam Site | 30 |
| 3.5 | Anderson Ranch Dam | 31 |
| 3.6 | Theron D. Weaver | 37 |
| 3.7 | Corps and Bureau "Spheres of Influence" | 39 |
| 3.8 | George H. Roderick | 41 |
| 3.9 | Mores Creek Bridge | 42 |
| 3.10 | "River of Gold" | 43 |

| | | |
|---|---|---|
| 4.1 | Lucky Peak outlet tunnel liner | 46 |
| 4.2 | "Rule Curves" | 49 |
| 4.3 | Plan of Lucky Peak Dam | 50 |
| 4.4 | A.H. Miller | 53 |
| 4.5 | Electrofishing the Boise River | 54 |
| 4.6 | Switching to Second Outlet Tunnel | 61 |
| 4.7 | William Ancell, Sen. Jim McClure, and Mayor Dick Eardley | 62 |
| 4.8 | West Boise Sewage Treatment Plant | 63 |
| 4.9 | Snotel Site | 64 |
| 4.10 | Auxiliary Outlet | 65 |
| 4.11 | Lucky Peak Dam and Power Plant | 66 |
| 5.1 | Summer rafters on the Boise River | 68 |
| 5.2 | Island forming in Boise River | 69 |
| 5.3 | Lander Street Sewage Treatment Plant | 69 |
| 5.4 | William Webb | 70 |
| 5.5 | Boating on Lucky Peak Reservoir | 71 |
| 5.6 | Boise Junior College | 72 |
| 5.7 | Ann Morrison Park | 73 |
| 5.8 | Ann Morrison Park vegetation | 73 |
| 5.9 | Boise River Greenbelt | 76 |
| 5.10 | Fairview Avenue Greenbelt underpass | 77 |
| 5.11 | Greenbelt Path near Shiloh Inn | 78 |
| 5.12 | Greenbelt Sign: Temporary End | 79 |
| 6.1 | Floodway schematic | 82 |
| 6.2 | Heron Lake at River Run | 88 |
| 6.3 | River Run water control works | 90 |
| 6.4 | Artificial stream at River Run | 92 |
| 6.5 | Greenbelt site near Boise downtown | 97 |
| 6.6 | Riverside Village entrance sign | 102 |
| 6.7 | Riverside Village "no trespassing" sign | 103 |
| 7.1 | Barber Park | 107 |
| 7.2 | Cottonwood trees in Boise River | 109 |
| 7.3 | Barber Park | 109 |
| 7.4 | ParkCenter Boulevard | 110 |
| 7.5 | Municipal Golf Course | 111 |

| 7.6 | Sandbags at Riverside Village | 112 |
| 7.7 | Floodwater downstream from Boise | 113 |
| 7.8 | Flooded bike path tunnel | 114 |
| 7.9 | Rip-rap | 115 |
| 7.10 | Garden City flooding | 116 |
| 7.11 | River Run levee | 117 |
| 8.1 | Development sign | 120 |
| 8.2 | Greenbelt underpass during flood | 126 |
| 8.3 | Revetment along Greenbelt | 127 |
| 8.4 | Lucky Peak "Rooster Tail" | 129 |

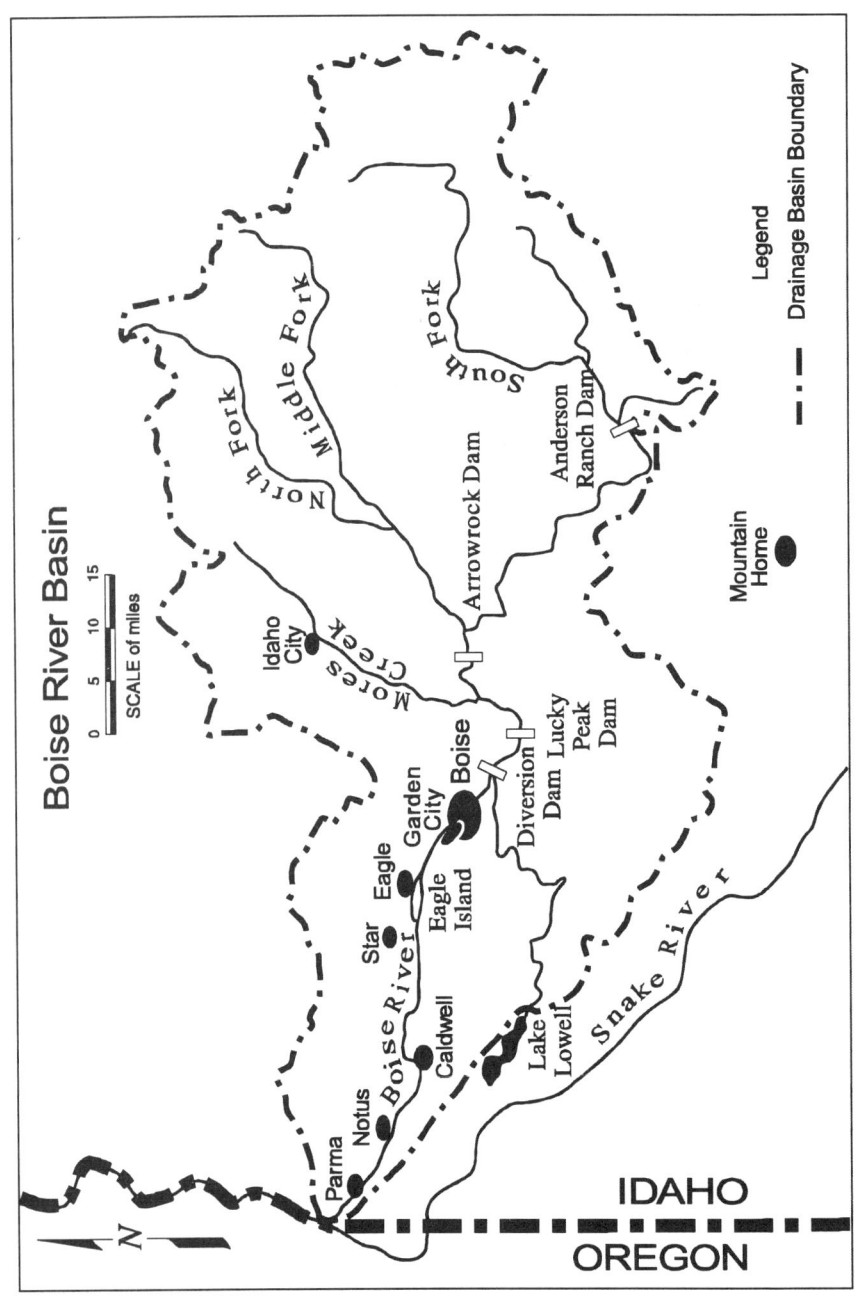

# Preface

Soon after I concluded a 13-year career in 1986 as a land-use planner with the city of Boise, of which seven years had been as director of the planning department, I undertook a contract for the Corps of Engineers to research and write a history of flood control on the Boise River. The Corps was particularly interested in the stretch of the river that flowed through Boise, the largest urban area through which the Boise River flows and the largest urban area within the jurisdiction of the Corps' Walla Walla District Office.

Before I began the work I thought that I knew a great deal about flood control on the Boise River. My staff had written flood plain zoning ordinances and plans, and we had processed all of the city applications for flood plain development. We had worked with the Corps' analysts in Walla Walla and frequently sought their advice and comment. However, as my research drew me into the files of other Boise city departments, into Idaho Fish and Game notebooks, into the old boxes that Walla Walla had dispatched to the Seattle Records Center, and into the other usual places that researchers go, I discovered how very little I knew. Life in the planning department had seemed so central, but I came to realize that the feeling of being central is part of what beguiles a bureaucrat. I began to imagine that each person stands in the center of a small circle where the view across the circumference can take in only a short segment of the arc. People generally look ahead and have little time or inclination to gaze at the past or to step into someone else's circle. Although I might have had the inclination, I had not the time. With the Corps contract, I had the time to discover a much bigger picture.

Because of this missing view of the big picture, participants in civic life have an excellent reason to write and study history. Had I and others understood more of the history of other players on the Boise River, and how certain patterns of action and reaction repeat themselves, perhaps we would have tried different strategies to achieve our goals. Indeed, perhaps the goals would have been different. Now hindsight—history—might provide someone else with foresight. The Boise River continues to be a lively and evolving focus of community development. It is very likely that agencies and interest groups will continue to play recognizable roles that originated long in the past.

I have been asked if I could write an "objective" history of a subject in which I played some part. Planners are obliged to sort out and assert the various "rights" of the public and property owners in controversial development proposals. Some property owners did not think that the public had any rights, while the public sometimes tried to deny property owners theirs. Planners try to find paths in the maze that lead to a balance of interests, usually via compromise. It should not be surprising then that the constant tension between private property and public interests is one of the chief lenses through which ex-planning directors see the world. The long-practiced habit of sorting out and articulating public interests in property development is not one I am likely to break. A career of public service provides me with a bias, the lens through which I see, and I disclose it here without apology.

# Acknowledgements

It is a pleasure to acknowledge the Office of History of the Corps of Engineers as the originator of this study of flood control on the Boise River. Dr. Martin Reuss administered the contract (DACW31–86–M–0775) with great attentiveness and respect for my independence in framing the analysis. Other employees of the Corps, particularly Ron Barrett and many of his colleagues at the Walla Walla District Office, extended many courtesies to me during my visits there, not the least of which was their willingness to talk candidly about the events of the past.

As I embarked on a masters degree program in history at Boise State University (BSU), the work took the form of a masters thesis under the guidance of Dr. Todd Shallat, director of the Public History Program at BSU, and Drs. Hugh Lovin and Glenn Barrett, the other members of my thesis committee. The history department, a wonderfully supportive environment for a student, awarded me a research fellowship, enabling me to do research and attend classes at the same time.

With book publication came a new set of associations—with editor, and now friend, Sylvia Dane at the Natural Hazards Research Applications and Information Center at the University of Colorado and Dr. Robert Sims, dean of the College of Social Sciences and Public Affairs at Boise State University. The two institutions co-published the book with support from the Corps of Engineers.

I owe much gratitude to everyone involved in bringing the work through these various incarnations. Each gave the manuscript critical evaluation, but gave me encouragement. During the research period, friends, former employees, and colleagues at the city of Boise welcomed me when I showed up to interview them or look at records. Thanks especially to Vickie Van Vliet, Judy Smith, Chuck Mickelson, Bill Ancell, Jack Cooper, and Tempra Wilson. They and many others made it easy to remember why working with them at the city of Boise was one of the best experiences of my life. To them and to all of the people who agreed to be interviewed, I give my thanks.

I owe a continuing debt to the librarians and archivists who have given me a great deal of individualized help, particularly at the Seattle Federal Records Center, Boise State University, the Idaho Historical Library, the Corps District Office in Walla Walla, Idaho Fish and Game, and the Record Center at Morrison-Knudsen Company in Boise. Special thanks to Elizabeth Jacox, Guila Ford, Garry Bettis, Alan Virta, Julie Kreiensieck, and all of the others.

Personal thanks go to Drs. Errol Jones and Patricia Ourada at BSU, Christina Stacy, editor Jeanette Germain, cartographer Lyman Larson, cover designer Pete Wilson, and book designer Gary Richardson. I must mention again my mentor Todd Shallat. Without him, the Corps contract would not have come about, nor the thesis or book. His aspirations for public history and his generosity to me and to all of his students provide a set of ideals worth striving for and a model of teaching worth imitating. I have always appreciated my husband's tolerance and patience, but never more than while working on a book, when those virtues get a real workout. To Ralph McAdams, my loving thanks.

December 1992<br>Boise, Idaho

## Before you start . . .

Here is a refresher on the terms measuring water quantity used in this book:

**Acre-feet**: a measure of the volume of water. One acre foot is the amount it would take to cover one acre of land with water one foot deep. The term is used to describe the capacity or current status of a reservoir. It is also used to describe the amount of water an irrigator might contract for delivery in a growing season.

**Cubic feet per second**: a measure of the amount of water passing a given place in one second. Abbreviated as "cfs."

The sketches on the next page are based on a pair of photographs made at Strawberry Glen Bridge early in the 1960s by William Webb of the Idaho Fish and Game Department. This bridge has since been removed. The first view is of the river in the spring when the flow of the river was estimated at about 5,300 cfs. The second view shows a late winter flow estimated at less than 200 cfs. One can imagine that a flow of 80 cfs is a thin shallow flow exposing even more of the bottom gravels and silt in the river channel. Around 2,000 cfs, the river is a pleasant stream, fun to float, and generally considered safe for such recreation. At 6,500 cfs, water fills the river channel, flows very swiftly, overflows the banks in some places, and is dangerous for rafting or floating. At 9,500 cfs today, there is a flood. Water through town is extremely high against the levees, overtops them in many places down river, and floods low ground throughout the river's course.

The flood of 1943 was about 21,000 cfs. Photographs show extensive inundation of farm areas, although the channel contained more of the flood than it would today. Some analysts, who base their estimates on the memories of the first Boise pioneers and on the record of known floods in other basins of the Pacific Northwest, think that the largest Boise River flood might have been 100,000 cfs in 1862, the year before Euro-American settlement began. The river would have spread from "bluff to bluff," covering the land between the foothills to the north of town and the old river terrace, locally called the Bench, to the south. See Appendix A.

Artist's sketch of the Boise River near the Strawberry Glen Bridge in June in the early 1960s, when the flow of the river was about 5,300 cfs.

Sketch of the same site with a late winter flow estimated at less than 200 cfs.

Idaho Historical Society, 72-95.6
**Diversion Dam and the Boise River canyon from the air, April 4, 1961. In the distance, Lucky Peak Reservoir and the mountains where the Boise finds its source.**

# Introduction

On the Boise River in southwestern Idaho sits a Corps of Engineers dam called Lucky Peak. When it was dedicated in 1955 a U.S. Army Corps of Engineers official said that the dam made the Boise one of the "most nearly perfect flood-controlled rivers in the country."[1] Two other dams, built primarily for irrigation, sit above it and contribute to the river's flood control system. Below Lucky Peak, the river flows through Boise, whose residents like to refer to it as "the beautiful Boise River," largely unaware of its exemplary status in the eyes of the Corps. They have little reason to, for the Boise River's pre-dam floods never did earn much of a reputation for wreaking damage or destruction on the city or the rural valley downstream. In a litany of historic American floods, one will hear of the Mississippi, the Ohio, the Susquehanna, the Columbia, the Trinity, scores of others—but never the Boise.

Since 1955 the Boise has become even more flood-controlled. In the 1970s land developers built levees, channels, and other flood-control devices along the banks of the river to protect residential and office developments in the flood plain. With technical assistance from the Corps, developers built against a theoretical 100-year flood. Not incidentally, they capitalized upon the investments that the public had made recently in the flood plain for a park system known as the Boise River Greenbelt. Since there was little development in the flood plain, these devices were not needed to protect buildings placed there by earlier inhabitants. Until Lucky Peak was built, and for the next 20 years, the Boise River flood plain was typically used for agriculture, gravel extraction, and certain industrial purposes.

Lucky Peak was one of hundreds of multiple-purpose flood-control dams that Congress authorized just after World War II. Martin Reuss, a Corps of Engineers historian, feels these dams probably would not have been built if flood control were their only benefit.[2] The advocates of Lucky Peak expected an irrigation project, far from the flood plain, to be a major beneficiary of the dam. In this respect, Lucky

Peak is not unusual, but reflects the national pattern of flood control projects at the time. It partly explains why the relatively modest floods of the Boise attracted the dam's "nearly perfect" flood control.

The wave of levees installed on the Boise in the 1970s, however, occurred during a sharp national debate on whether such structures were an effective means of limiting the cost of flood damage over the long run.[3] The "engineered structure" approach to flood control came under attack after 1966 when the Task Force on Federal Flood Control Policy reported to President Lyndon Johnson that, despite the billions of dollars expended in dams and levees, the annual bill for flood damage continued to grow.[4] Structural control did not prevent rising damage costs as land development increased and inflation escalated. In the aftermath of the report, policymakers started serious discussion of nonstructural methods such as land-use management, emergency preparedness, flood-proofing, and removal of buildings vulnerable to flooding.[5]

The debate and criticism of structural controls led Congress in 1974 to enact a provision in the Water Resources Development Act requiring federal agencies to consider nonstructural alternatives for flood-damage prevention in their planning process.[6] The Corps of Engineers instructed all of its district offices to "give thorough treatment to nonstructural plans" and to prepare nonstructural flood plain alternatives on the same level of detail as structural alternatives.[7] The Corps implemented or advocated nonstructural approaches in Arizona, Wisconsin, Massachusetts, and Colorado. These projects enjoyed the praise of environmental and mainstream presses alike. The *New York Times* lauded Chief of Engineers General Frederick J. Clarke for saying, "We can't guarantee that no place will get flooded. People have to expect it every so often." We need laws, he said, "to prevent encroachment on the flood plain of our streams."[8]

This national sentiment, or at least any effective implementation of it, seemed to pass unnoticed by Boise and the Walla Walla District of the U.S. Army Corps of Engineers. In Boise the local zoning laws were changed so that the flood plain could be narrowed by new levees and other "structural controls." Houses and buildings were contructed on land thus removed from the 100-year flood danger. In an effort to better understand these events, this book examines the history of flood control on the Boise and the relationship between federal policy and local response.

The tale of economic development via flood control on the Boise River, while adding a chapter to Boise's history, is also a chapter in a larger historical tapestry— the politics of western water. The Boise Valley is part of the arid West, with annual average precipitation of 11 to 12 inches. It is the site of one of the early reclamation projects undertaken by the Reclamation Service (now known as the Bureau of Reclamation) after its creation in 1902. As such, it is one of the places in historian Donald Worster's "hydraulic west," characterized by "faceless, anonymous, impersonal" dams, a land of canals, storage reservoirs, and ditch riders.[9] An air flight over Boise Valley reveals the hard edge where a ribbon-like canal marks the end of the desert and the beginning of lush vegetation and the contours of the plow. Less visible—but present—are the weirs, ditches, culverts, siphons, pumps, and headgates that make the water delivery system work. Like settlers elsewhere in the West, Boise Valley farmers learned how to claim, move, manipulate, save, and guard their water—primarily for irrigated agriculture.

The agricultural claim on water in the Boise Valley went largely unchallenged through the first half of this century. After domestic use, irrigation was dominant, other uses secondary. Public discussions about water centered on ways of getting more of it for more irrigation, rarely on fish habitat, wetlands, sewage dilution, or riparian aesthetics. After 1955, when Lucky Peak went on line, the situation changed significantly. The population grew more dense, urban growth subdivided farm fields, the economy diversified beyond dairying and agriculture, and people enjoyed more leisure time. Then the environmental movement rearranged public opinion so that it became respectable to protect eagle habitat instead of promoting human habitat.

The Corps official speaking in 1955 might as well have said that the Boise River was nearly under the total dominion of human society with little left for natural forces to decide. This book considers how and why this happened, which interests in society benefited under that dominion and which did not, and whether the beneficiaries in 1955 were the same as the beneficiaries 30 years later. Much has changed. The question is to determine the nature and meaning of this change, however slowly it may have occurred. In the day-to-day management of the Boise River's water, originally designed to serve irrigation and flood control, other water users have managed to wedge ever-stronger claims on this western river. The early Idaho consensus about irrigation, what Idaho lawyer Scott Reed called "legal water-right apartheid," has, at the least, begun to break down.[10]

The Corps of Engineers has exerted a major influence on the Boise River. What has been its role in determining who benefits from how the river is managed? Hundreds of books have examined the Corps of Engineers and its relationship with various interest groups. Some have been judgmental—casting the agency as arrogant, irresponsible, an environmental pillager. Some have defended it against these attacks.[11] Others see in the agency a convenient social science laboratory, a place to test theories about decision making, change, and resistance to change.[12] Books about the Corps often inspect its relationship to Congress, the Bureau of Reclamation, or to national pressure groups. Amid this mass of books, this history places the Corps vis-a-vis "local interests," reminding us of former Speaker of the House Tip O'Neal's famous axiom: "All politics are local."[13]

# Chapter One
# When Were the Floods?

James Brown, editor of the *Idaho Daily Statesman*, was shocked when he realized in 1953 that the U.S. Army Corps of Engineers was spending $20 million to build Lucky Peak Dam on the Boise River. "When were the floods?" he demanded.[1] He could not remember any flood serious enough to call for such extravagant federal expense. But his protest was too late; the dam already was half finished. Brown mistakenly told his readers that the dam "will flood out Arrowrock Dam," a Bureau of Reclamation irrigation storage dam just above Lucky Peak. He worried, also mistakenly, that the project might somehow cause water levels around Meridian, a town about 12 miles west of Boise, to rise. Nevertheless, regarding Boise Valley's flood history, he had a point. Flooding alone never had done enough damage to justify the dam. Brown's editorial (and the one that followed two weeks later to correct his facts about Arrowrock being inundated) shows that he had no idea why Congress had in fact decided that Lucky Peak was worth the money.

The Boise River rises every spring, carrying snow melt away from the spectacular glacier-carved granite peaks of the Sawtooth Mountains. Descending in a westerly direction, the river emerges from its mountainous environment through a narrow canyon and flows through the gradually widening Boise Valley for about 60 miles until it joins the Snake River. The Snake joins the Columbia, and the Columbia empties into the Pacific Ocean at Portland, Oregon. Some years the Boise brought more water, others less. Boise Valley settlers learned to expect this rising river and the shallow flooding it often brought. As the settlers grew more experienced with the river, they managed to defend themselves against the inconvenience of floods and later to put the water to productive use.

Between 1863 when settlement began and 1946 when Congress authorized Lucky Peak Dam, there were many occasions when the community referred to the spring freshet as a "flood." Three of these, the floods of 1876, 1896, and 1943, can illustrate the valley's development, its ever-expanding use of irrigation, the changing use it made of urban and rural flood-plain lands, and its increasingly complex response to flood problems. Naturally, the community's growth affected the magnitude and type of damage caused by flooding.

The first of these, the flood of 1876, occurred during an average runoff year, and is useful to contrast with the later larger floods on the Boise. A detailed account of this flood is available because the *Statesman* editor at the time was traveling — by horse — to cover another story and described the high water and other impacts of the flood as he observed them. The 1896 flood was remembered by witnesses still living in 1944 when they described it in a public hearing held to discuss the Lucky Peak Dam proposal. That flood was the last big flood before the Reclamation Service constructed Arrowrock Dam, the first of the storage dams built on the Boise. The river produced its largest 20th-century flood in 1943, a few years after the second storage reservoir, Anderson Ranch, was authorized, but before it was finished. The 1943 flood eventually led to the congressional authorization of Lucky Peak Dam.

### The 1876 Flood: Early Settlement

Euro-American settlers arriving in Boise Valley in 1863 watched the river, a braided stream with wide meanders and interlacing channels that divided and rejoined around substantial islands, swell every spring, like most other rivers that rise in the high mountains of the American West. At times the water escaped the banks, the severity of the flood depending upon the depth of the mountain snow, its moisture content, and the rise and fall of spring temperatures. Typically, more than two-thirds of the annual discharge of the river occurred in May.[2] Most of the spring snowmelt escaped agricultural use, and the settlers ruefully regarded it as wasted, for the climate was dry indeed, with annual precipitation amounting only to 11 or 12 inches in a year.

In July of 1863 a handful of pioneers platted the town of Boise and its Main Street about three-fourths of a mile north of the river. They planned for orchards and farms between the town and the river, aware of the potential for diverting water from the river for irrigation.[3] The location of the flood plain was undoubtedly obvious because a heavy growth of cottonwood trees and willows defined its edge. Tom Davis, one of these first city planners, had diverted part of the river flow and begun his irrigation system earlier that spring.[4] The population of the valley steadily increased, its economy based upon supplying fresh food and commercial services to the mines of nearby Idaho City and the military garrison established in Boise in 1863.

Other farmers and orchardists immediately began doing the same type of irrigation as Tom Davis. They settled the entire 60-mile length of the valley within

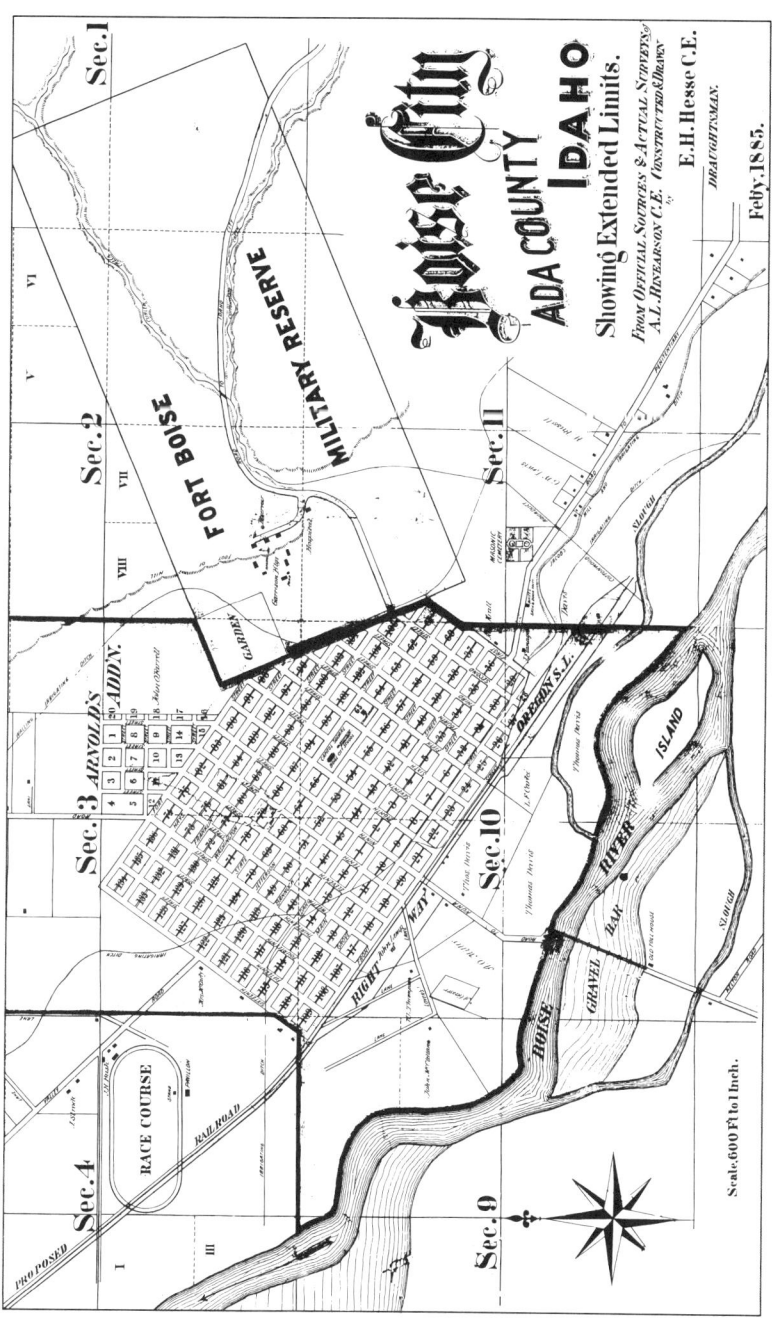

Figure 1.1
An 1885 plat shows the distance between town and river. The Ninth Street Bridge carried the "Road to Kuna" across the main channel of the river, a gravel bar, and a "slough," later called the South Channel.

the next few years. By 1876, over 50 individuals and ditch companies had recorded water rights claims on the Boise.[5] Depending on the local topography and the amount of land to be irrigated by these gravity-flow diversions, farms filled in three to five miles on each side of the river.

Judge Milton Kelly, the editor of the *Statesman*, decided one spring day in 1876 to ride the entire length of the valley and report on how these farmers were doing.[6] The river had begun to rise against its banks a few days before he departed, and a few reports of damage already had arrived. Water had gone over the banks at two or three farms, inundated an orchard, and floated lumber away from one of the lumber mills above the city. The river had cut a new channel and left the old one nearly dry. As a rule, spring floods overwhelmed the farmers' crude water diversion works, and this year was no different.[7]

Despite the freshet, Kelly mounted his horse and headed down river. One of his first stops was at "Aiken's farm," a two-part establishment consisting of 360 upland acres and another 160 acres closer to the river. Aiken had just dug a ditch from the river and finished a distribution system for the 160-acre parcel. As Kelly rode on down river, he noted that the most easily irrigated ground was settled, but that there were still a few parcels left for future development. Ever a promoter, Kelly hoped that a few "go ahead men" like Aiken would soon get busy with this last work.[8]

*Idaho State Historical Society #345-B*
**Figure 1.2**
**Milton Kelly, editor of the *Idaho Daily Statesman*, described the 1876 flood.**

Farmers "have a good deal of fear of high water," Kelly wrote, "but have taken great pains to levy and guard against it."[9] One farm required more than a mile of levee. Since the flooding was rather shallow, the levee was only one or two feet high and not very costly to build. Maintenance, however, was another matter— a constant skirmish with gophers that burrowed endlessly, allowing flood waters to enter the tunnels and erode the levee. The solitary farmer had to be vigilant to keep the levees intact and functioning during a flood.[10]

The practice of building inexpensive levees and then defending them in a flood fight continued to be standard procedure for the next 80 years, even though it was not always an effective method. As the river coursed through the valley, it fell 500 feet in its 60-mile trip to the

Snake River, an average fall of 12 feet per mile. This drop was enough to generate considerable velocity when water ran high. Since the farmers made their levees with the local gravels of the valley floor, time and exposure caused them to erode and break up. Heavier materials could be used, and sometimes were, but it was not always worth the expense, since not every year brought a bad flood.

As Kelly progressed closer to the mouth of the river on his 1876 ride, he observed that some farmers had moved off their places. One man had two feet of water in his house. Despite these examples, Kelly decided that reports of ruined farms had been exaggerated. He perceived flooding as beneficial to soils and growing crops. His optimism might have been somewhat calculated, however, for his editorial mission was to encourage the continued settlement and economic development of the Boise Valley. Too harsh an estimate of conditions would not have suited such a purpose.[11]

While the farmers were busy with levee work, residents of Boise tried to save the Ninth Street Bridge, the only one serving the town. The bridge was actually two bridges connecting two channels of the river over an island. Workers threw rock and stone onto one of the abutments, hoping to increase its resistance to erosion. The effort failed when the wave action of the river was too much for the rock; the apron and one span of the bridge were lost. As flood water advanced up the approach road, the Chinese farmers who worked nearby made a temporary dam and successfully protected their vegetable gardens.[12]

The urban and agricultural parts of the valley experienced flood damage very differently. The town of Boise was still small and confined to the north side of the river. No urban property was damaged except the bridge, which cost $6,000 to repair.[13] The flood plain was used almost entirely for farming. Agricultural damage consisted of erosion, loss of topsoil, deposition of debris, inundation, and loss of fencing. Farmers worked independently in their flood fights because they lived fairly far apart from one another and had their own particular flood worries. Lumber mill owners, whose operations were right on the river because they floated logs down from the forests above, usually anticipated some degree of annual flooding and managed accordingly.

Judge Kelly was in a stoic mood after the flood and urged that those in the flood danger area should plan ahead, build good levees, and keep them maintained. There was snow in the mountains after all, and "Boise River furnishes one of its modes of egress and passage to the sea, so we shall have to let it come down."[14] The Corps of Engineers later estimated that the peak discharge for this flood was 15,200 cubic feet per second (cfs), a fairly normal flood during the years before there were any dams on the Boise.[15]

## The 1896 Flood: Irrigation Expands

Twenty years later in 1896, the peak spring runoff was 35,500 cfs, more than twice the volume in 1876. This was not an estimate, as measuring devices had been installed in the river the year before.[16] Land use in the flood plain was now more intense, though it was still mostly agricultural. This time there were no reports of standing water in houses. Perhaps farmers used more sophistication in the location of their homes now that they had 20 more years of experience with the river. Some of the early orchards still stood, but many riverside land owners near Boise leased out to truck gardeners who provided fruit and vegetables for city residents.

The city now had its second bridge, the Broadway, located about a mile upstream from Ninth Street. Further down river, irrigation farming in the valley continued much as before. New settlers had by now filed claims for more water than the natural flow of the river could ever satisfy. Since one of the principles of western water law is "first in time, first in right," settlers with late claims were out of luck if those with prior rights used the water. Such farmers often lost their investments in low-water years when their crops parched.

What was new since 1876 was the diversion of water to the Bench, the desert lands above the valley to the south. The first such irrigation began in 1878 when an investor named William B. Morris built the Ridenbaugh Canal.[17] Latecomers to the Ridenbaugh and other projects had marginal operations because of the unreliable water supply in late summer. They suffered severe loss during low-water years. As each spring saw the wasting of water on its way to the sea and each summer drought took its toll, developers and community builders knew that they must have storage reservoirs if they were to have enough water.

East of town, lumber establishments dominated the flood plain, while nearer the business district, warehouses and utilities began using some of the land between the

**Figure 1.3**  *Idaho State Historical Society #81-32.10/e*
**Sawmill operators above Boise sometimes lost control of the logs in their booms.**

river and Main Street. Housing had crept closer to the river and occupied the Central Addition, an area known today as the Myrtle Street neighborhood. A developer had platted a small residential subdivision on the south side of the river, no doubt inspired by the construction of the Broadway Bridge. People who lived there tended to move their possessions to higher ground whenever water breached the banks or levees upstream of the bridge, but few seem to have been inundated.

The *Statesman* began covering the flood on May 14, 1896, and continued until about June 17. The water rose and fell several times depending on how temperature changes affected the rate of snowmelt. The current was so strong it sucked logs under the boom at the Rossi lumber mill and carried them down towards town. They lodged against the trestlework and piers of the Broadway Bridge and acted as battering rams. The sheriff kept crews there to saw them up and clear them away. Other escaped logs eventually beached themselves where crews piled them atop one another in the hope that their weight would help bulwark the banks.[18]

The fight for Broadway Bridge aimed to prevent the river from cutting behind the abutments to create a new channel across the approach roads. (In future years, a better understanding of this type of threat inspired the straightening and clearing of the channel just upstream of new bridges as a protective measure. Bridges not supported by bedrock were vulnerable to changes in direction of the river current as well as high velocities.) The battle took on a human dimension when a group of Central Addition residents concluded that if the river were left alone to cut behind Broadway Bridge on the south side, it would reduce the pressure on their own north side. They appealed to the chair of the Ada County Commission to let the bridge go. He refused, declaring that access across the bridge was more important than possible damage to their homes. Thereafter he stationed a deputy at the bridge with a Winchester rifle just to make sure. The incident illustrates the extent to which

**Figure 1.4**  *Idaho State Historical Society #2045-B*
**Built in 1892, the Broadway Bridge was in jeopardy during the flood of 1896.**

public management of the flood had become a practice. In the end the Central Addition was not flooded. But seepage was bad enough to mire the ice wagons traveling their routes on the dirt streets of the neighborhood in mud.[19]

Further down river in adjacent Canyon County at Caldwell, a town smaller than Boise, residents criticized county officials for acting too slowly in the crisis. The headgates of a flooded ditch gave way and cut the town off from the main wagon bridge. Someone threw a footbridge over the channel and restored contact. The commissioners finally organized flood crews after a road washed out.[20]

When the river subsided, the biggest urban victims of flooding were once again the bridges across the river. Farmers who practiced row cropping in their bottom lands, as the Chinese truck gardeners did, ran the most risk of losing their acreage if their levees failed. At least one barn built too close to the river was lost. Someone noticed it "floating serenely off toward Caldwell."[21]

Thirty-three years after the city's founding, and 20 years after the flood of 1876, the community had more people to bring to bear during a crisis and was more organized to make collective decisions about how to respond to a flood. Levees continued to be the major defense. Flood-plain land was still mostly agricultural, and the river now irrigated bench lands above the valley. One thing that had not changed was the deep regret with which farmers watched the flood waters flow uselessly out of control to the ocean.

## The 1943 Flood: Federal Reclamation

By 1943 storage reservoirs were at last a reality. Agriculture in the Boise Valley had significantly expanded beyond its 1896 limits. The early hopes of "making the desert bloom," as civic boosters liked to put it, had finally materialized with the intervention of the federal government. The Newlands Reclamation Act of 1902 set up a procedure, an agency, and a method of financing large-scale irrigation projects. In 1909 the Reclamation Service completed the Boise Project, which consisted of Diversion Dam about seven miles above Boise City; the 40-mile New York Canal; and Deer Flat, an off-river storage reservoir now called Lake Lowell.[22] At last, irrigators could divert part of the spring flood into the reservoir and release it from storage later in the summer.

The 177,000 acre-feet of storage capacity at Deer Flat Reservoir was not enough to satisfy settlers' needs, so the Reclamation Service built a second reservoir; Arrowrock was on the river about 18 miles above the New York Canal Diversion. At 350 feet, Arrowrock Dam was the highest dam in the world when completed in 1915. It stored another 286,000 acre-feet for irrigation. The engineers included a special log chute at the dam that could transfer a million board feet of logs a day from the reservoir to the lumber mills downstream.[23]

The two storage reservoirs made a big difference in the outlook for the Boise Valley. In 1900 there had been insufficient water to support the valley's 1,600 farms; retrenchment and a decline in population seemed likely. Twenty years later the number of farms had tripled; there were 374,218 acres under the plow.[24] Then

**Figure 1.5**  *Idaho State Historical Society #2652*
**Diversion Dam, dedicated in 1909, sent part of the river into the New York Canal, the structure to the right in the picture.**

**Figure 1.6**  *Idaho State Historical Society 70-10.1224*
**Arrowrock Dam in 1916. Water flows over the spillway on left. Note log chute on far right.**

in the 1920s came a discouraging period of drought. In a series of low-water years, the reservoirs were unable to supply enough water to those irrigators with the most recently acquired water rights. Combined with low crop prices in the 1920s, the drought meant disaster for many.

Irrigators were organized into irrigation districts, which were single-purpose governing bodies that each had its own elected board of directors and the power to collect revenues and establish policy. Five of these districts organized themselves as the Boise Project Board of Control in 1927. This board took over from the Bureau of Reclamation (formerly named the Reclamation Service) the responsibility for collecting water-use fees from individual farmers and for managing the water distribution system in the project. In 1937 the Bureau added five feet to the crest of Arrowrock Dam, providing storage for another 30,000 acre-feet.[25] It was still not enough. In 1940 the Bureau started Anderson Ranch Dam, a huge structure 42 miles upstream of Arrowrock on the South Fork of the Boise. It was designed to hold 500,000 acre-feet, which was enough to end drought insecurities in all but one year out of 20.[26]

Anderson Ranch was a multiple-purpose reservoir. The top 212,500 acre-feet of storage capacity were set aside for flood waters. Preliminary studies had indicated that the reservoir would satisfy both irrigation and flood-control needs. The Bureau and Corps analysts considered the channel capacity in the Boise Valley to be 10,000 cfs, and the new reservoir was designed to control floods to that flow.[27] With the outbreak of World War II, however, the War Production Board ordered most of the equipment and labor at the dam site put to other uses. Unfinished Anderson Ranch played no role in mitigating the 1943 flood.

That flood was to visit a town and valley vastly changed since the 1890s. The increase in irrigated acreage helped propel the population in the Boise Valley from 10,000 in 1890 to 95,000 in 1940. Boise became the largest town along the river with 26,130 people in 1940. As in many parts of the arid West, urban growth was a side effect of irrigation development.[28] Boise's physical expansion was blocked on the north by the foothills of the Boise Front, so the city jumped across the river to the south, and also spread east and west. In addition, some land uses became situated closer to the river. Local government began to fill lowlands and straighten the river channel to protect bridges, particularly below Ninth Street.[29]

In the 1930s, Great Depression employment programs and federal flood policy each served to enhance flood control along the Boise. After severe floods on the Ohio River, Congress in 1936 declared flood control to be a national issue and gave the Corps of Engineers a nationwide flood-control mission and budget.[30] Boise Valley farmers secured a modest share of these funds to improve their levees. The Idaho Legislature responded to a Boise River flood in 1936 by appropriating $10,000 for work to confine the channel. The state highway department and the Works Progress Administration added more funds and labor for levee repair.[31]

A few riverfront farms now gave way to food processing plants that used the river to dispose of their untreated wastes. Eighth and Ninth streets filled up with warehouses and manufacturing establishments all the way to the bridge. A number of gravel extraction operations mined the flood plain for its high-quality river rock.

After 1940 farmers on many riverfront farms west of Boise changed the way they farmed their flood-plain lands. They started planting row crops. But when they switched over to row cropping, they did not necessarily improve their flood defenses. C. Ben Ross, one of Idaho's governors and a Boise Valley farmer, once spoke of the way prudent farmers had learned over the years to keep their flood-prone fields in pasture, planting grasses, shrubs, and trees. When the floods came, the soil resisted erosion, held in place by tight root systems. In this way the land continued to be productive, if not lucrative. To plow up floodable land was to risk its complete loss. After the United States entered World War II, however, farm produce prices rose dramatically. Farmers, encouraged by the government, tempted by high market prices, and inspired by patriotism, took the risk and planted their low-lying lands to row crops. They harvested excellent yields and commensurate profits.[32] Then came the 1943 flood.

Arrowrock and Deer Flat reservoirs mitigated the flooding on the Boise in most years. The annual flow volume of the river ranged between 890,000 acre-feet in a low snow year to a maximum of 3.57 million acre-feet in a high one. The average

**Figure 1.7** *U.S. Army Corps of Engineers*
**Aerial view of Plantation Golf Course, April 21, 1943. A failed levee allowed water out of the river channel. State Street in lower part of picture.**

was 1.8 million acre-feet. The two reservoirs held 463,500 acre-feet, about a quarter of the average yearly flow. In moderate runoff years the dam caught the spring gush and reduced the impact of the flood. Occasionally, nature sent snowmelt into the reservoir faster than its 10,000 cfs capacity discharge ports could let it out. It was just a matter of waiting for the reservoir to fill up before water went over the spillway. After that, whatever entered the reservoir flowed out as though the dam were not there. Some could be diverted to Deer Flat, but the canal could only take about 2,820 cfs. The Corps estimated that between 1915 and 1943 Arrowrock reduced the spring high water by an average of 3,443 cfs each year. Such buffering had a less welcome side effect of reducing river channel capacity.[33]

The Arrowrock buffer was not enough to prevent a damaging flood in 1943. An unseasonable warm spell in early April accelerated the snowmelt. Boise River watermaster William Welsh warned Governor C.A. Bottolfsen that without a drastic change in the weather the water behind the dam would tumble uncontrolled over the spillway in another 15 days.[34] Considering the substantial snowpack, Welsh anticipated the flooding of 500 farms between Boise and the Snake River.

Welsh mobilized the resources in the community and activated plans providing for typhoid inoculations, evacuations, and emergency levee repair. Property owners in the flood plain prepared to do as their predecessors had done: stand their ground, guard the levees, and fight the flood. This time, however, there was a two-week notice, modern communications, a supply of U.S. Army troops, and highly centralized professional leadership. Welsh organized the river into eight districts, each with committees to deal with communications, evacuations, and emergency loans. Since most residents had radios but not telephones, organizers decided to broadcast flood information at 11:30 a.m. and 12:30 p.m. daily while farmers were in from the fields at their midday meal.[35]

The principles of levee reinforcement were the same as they had always been, but now bulldozers and other heavy equipment supplemented the work. In addition to stones and escaped timbers, rip-rap material now included the 20th-century innovation of automobile bodies filled with rock.[36] Military detachments stationed at Gowen Air Field near Boise provided the kind of concentrated labor

*Idaho State Historical Society #3535*
**Figure 1.8**
**William Welsh, Boise River Watermaster during the 1943 flood.**

13

**Figure 1.9**
*U.S. Army Corps of Engineers*
**Aerial view on April 21, 1943, shows town of Eagle in lower part of picture and Eagle Road with washed out culvert.**

**Figure 1.10**
*U.S. Army Corps of Engineers*
**The Linder Bridge was one of the many swamped by the 1943 flood.**

needed at bridges, the Chinese gardens, and other sites down the valley. The Corps of Engineers supervised construction work at levees.

At its peak, about 20,000 cfs churned down the river.[37] Islands, eroded banks, and lowlands were inundated as usual. Listings of threatened places included a number of new land uses in the flood plain: the grounds of the Old Soldiers Home at Veterans Park, the access road to the Idaho Fish and Game Hatchery, a tavern south of the Broadway Bridge, and parts of the Plantation Golf Course and the Triangle K construction and scrap yard.[38] Near the mouth of the river, several houses took in water above the first floor amidst rumors that this particular flooding had occurred because someone had deliberately blown up a ditch.[39] In total, 200 families and their livestock evacuated their farms while Idaho Volunteer Reserves guarded them against pilfering.[40]

Damage on lands where levees had failed (or where there were no levees) included loss of topsoil, deposits of sand and gravel, potholes, loss of rip-rap and dike work, washed out culverts, damaged ditches, piles of debris, and loss of crop yield at harvest. Landowners had to move livestock and buy feed, then reseed and relevel fields later.[41]

Bridge failures constituted costly damage. Not counting the railroad bridges, there were 14 highway bridges across the Boise River. Authorities shut down all but three for five days or more during the flood emergency.[42] The bridges' chief weaknesses were unstable piers and foundations. At the Fairview Avenue Bridge, crews cut a hole in the surface of the bridge through which heavy trucks dumped ton

**Figure 1.11**  *U.S. Army Corps of Engineers*
**Plantation Golf Course on April 22, 1943.**

after ton of lava rock hauled from the Boise Canyon. The rock bulwarked the south abutment, where wave action and current battered and weakened it.

With bridges closed, flood damage now included losses by dairies and other businesses that had extra costs for out-of-way travel to a functioning bridge.[43] The proprietors of establishments near the river took preventive measures such as moving goods out of basements and sandbagging the grounds, and apparently avoided flood losses. Several house basements flooded in the Broadway Avenue area south of the river, but this flooding was caused by sewers that took in water at their riverbank outlets. These were homes where residents failed to carve wooden plugs and stuff them into their sewer pipes as local plumbers advised.[44]

The Corps of Engineers' first impression of the flood was that "only a very small amount of damage has been done in Boise Valley. The fact damage has been held to a minimum is . . . due to excellent organizational work of you [Gov. Bottolfsen] and your associates."[45] This appraisal seemed premature and rather brief to Idaho Senator Henry Dworshak. He asked Corps Portland District Engineer Oliver Lewis to perform a damage survey.

The Corps published a summary of the information from the survey in its 1946 proposal to Congress to build Lucky Peak Dam.[46] This report also contained a transcript of a 1944 public hearing the Corps conducted about the flood.[47]

The Corps tallied the damage as follows:[48]

| Type of Damage | Amount in $ | Percent of Total Damage |
| --- | --- | --- |
| Agriculture | $649,710 | 65.1 % |
| Urban/Suburban | 0 | 0.0 |
| Industrial | 18,640 | 1.8 |
| Roads/Bridges | 76,040 | 7.6 |
| Utilities | 200 | 0.0 |
| Railroads | 0 | 0.0 |
| Levees | 205,440 | 20.6 |
| Emergency activity | 18,220 | 1.8 |
| Indirect | 29,100 | 3.0 |
| TOTAL | $997,350 | 99.9 % |

The testimony contained details that, while differing somewhat from the Corps' generalized figures, provide additional insight into the nature of the flood damage. Mayor Austin Walker, the only Boise witness among the 87 who testified, had asked the city's households and businesses to send him a list of their flood-related damage. These totaled $6,769 for private costs, of which a small portion was submitted by utilities. State and county highway officials reported that bridge and road repair costs were in excess of $200,000, somewhat higher than the Corps' reported $76,040. Repair of the Star Bridge alone, where a pier had settled two feet, was accounted at $104,500. The state spent at least $7,000 to protect Fairview Bridge during the flood and then $24,724 to raise the one pier that had settled.

The "indirect" damage of $29,100 consisted of interruptions in traffic and communications due to the bridge closures. The Corps particularly noted the handicap to large dairy creameries. In addition, the flood temporarily interfered with the delivery of water to irrigated lands beyond the flood plain.

In summary, Boise Valley managed the 1943 flood fight with no loss of life and little structural damage in the city except for the bridges. The rural areas absorbed most of the damage, with two-thirds of the $1 million in losses spread across hundreds of flood-plain farms.

If there had been a flood-control dam above the Boise Valley, most or all of the 1943 flood damage might have been prevented. Naturally, the damage that could be predicted for the years to come would also be prevented. But, as Corps benefit/cost analysis would soon show, all of the benefits of avoiding future damage would prove to be less than the cost of building and operating the dam. If a dam was to be built, other benefits would have to be added to the benefit/cost equation.

**Figure 1.12** *U.S. Army Corps of Engineers*
**View to the west (downstream) on April 22, 1943. The flooded areas in the foreground and center, later protected by levees, became the site of the River Run housing development and ParkCenter office building complex. At upper center of picture, just beyond the Broadway Bridge, the river remained within its banks behind a levee on left side and the railroad right-of-way and fill on right. Olson City Steel, later Gate City Steel, at the end of Warm Springs Avenue can be seen at lower right.**

# Chapter Two
# After the Flood

In the muddy aftermath of the 1943 flood, the seeming consensus of public discussion was that something had to be done to prevent another one like it. But it was not clear that the solution to the flood problem would be a large dam. Several different ideas emerged in the discussions that took place in Boise in the summer and fall of 1943. Only one of them evoked a response at a federal level.

### Local Responses

One proposed solution came from the Ada County Board of Commissioners before the flood had receded. The board proposed a $3-million postwar project to widen and straighten the Boise River channel. Commissioner J.M. Dobbs suggested that both Ada and Canyon counties could participate together with federal assistance to build new permanent dikes, widen the river, and clear the channel of debris and sandbars.[1]

Watermaster William Welsh, however, began urging two other kinds of help from the government. He wanted immediate assistance from the U.S. Army Corps of Engineers to repair key levees along the river, and he wanted the Corps to restudy the flood problems of the river. Anderson Ranch could not be in service for flood control for at least two or three years, he thought, even if the War Production Board allowed construction to resume. He believed that the flood had decreased the carrying capacity of the river to somewhere between 7,000 to 8,000 cfs, and that the Anderson Ranch solution might not be sufficient to guarantee flood protection given this changed condition. Besides, there was another potential use for flood waters. In May 1943, Welsh wrote Portland District Engineer Colonel Donald J. Leehey, "I do hope that someday a way may be found to store this flood water, thus not only preventing these disastrous floods, but also making the water serve a useful purpose for irrigation and power development."[2]

A historical analysis by State Reclamation Engineer Mark Kulp generated another approach. Kulp compared two past floods, those of 1936 and 1938; each had been approximately 19,000 cfs. After the first one, the legislature had voted $10,000 to improve the river channel, modify some of the bridges, and clean out timber and debris from the river and its adjacent banks. (During the debate over that appropriation, a skeptical Canyon County legislator remarked, "If there ever was a river where a duck had to pack a water bag, it's the Boise."[3]) The 1938 flood then came and did no appreciable damage.[4] Kulp felt, therefore, that diligent stream maintenance was an essential and effective part of flood protection. He noted that in-stream diversion structures slowed the flow of water and caused increased sediment deposition in the river channel. Channel maintenance had to become a routine feature of annual water diversions for irrigation.

Another proposed solution looked to the source of the siltation that was reducing the channel capacity in the river. A farmer, Robert M. White, accused the federal government of negligence because the old Territorial Legislature had allowed mine tailings to be washed into the river. He declared that the reason gravel bars in the valley got so high was because they consisted of the mountains

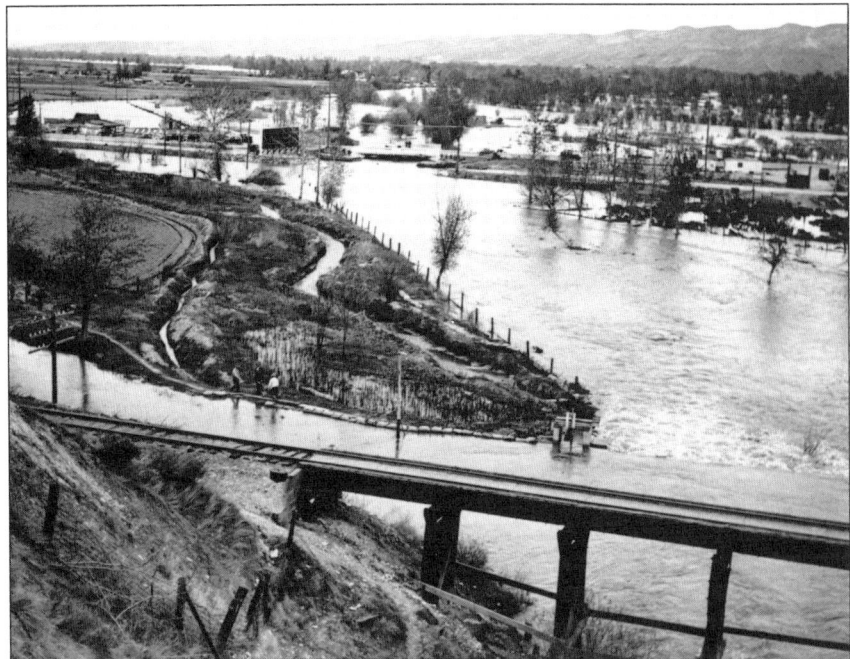

**Figure 2.1**  *Idaho State Historical Society #68-23.5*
**Flood scene in 1936, showing Chinese garden to the left and sandbag work. After this flood, the Idaho legislature spent $10,000 to remove debris and improve the channel.**

themselves, eroded and carried down by runoff. Mores Creek, one of the main tributaries to the Boise, whose headwaters provided hydraulic force for placer mining, was infamous for its silty waters. His thoughts foreshadowed later recommendations from the Department of Agriculture's Soil Conservation Service for better watershed management as one way to prevent flooding.[5]

John Kent, an engineer with the Bureau of Reclamation, suggested that the Boise be re-formed with a broad, flat river bottom with high banks and a channel not so deep that it would interfere with the irrigation canals during low-water periods.[6] With the exception of "high banks," by which he undoubtedly meant levees, this proposal was most similar to the natural condition of the Boise River flood plain. Not long after Lucky Peak was built, Corps officials would propose just such a coordinated levee system.

Valley farmers with land near the river simply hoped the federal government could offer them relief from flooding. Damage was a chronic nuisance, even though the seriousness of the nuisance varied from year to year. They had enjoyed the help thus far provided by the Corps for levee improvements, but as a group, the farmers did not sponsor an organized proposal. Their suggestions were based on their understanding of the past. They agreed that the river needed to be maintained, cleaned of debris, straightened out at some of its bends, made wider and in some places deeper, but not so deep that it would interfere with their diversion structures.

**Figure 2.2**  *U.S. Army Corps of Engineers*
**Farmers had to bear costs of moving their livestock out of the way of the flood, then of removing debris, releveling and replanting. Boise River is beyond the trees in the background. This alfalfa field is seven miles below Boise.**

The farmers who relied on levees knew that better levees could be built. They called for assistance in constructing revetments and other improvements that would outlast the kind of works they had built themselves.[7] Some recognized the flood protection benefit of upstream reservoirs and several suggested either an additional reservoir or a series of small ones. A few recalled the unusual contribution to the 1943 flood from Mores Creek, one of the four main tributaries to the Boise, and suggested a Mores Creek reservoir.

It is not surprising that the flood-plain victims called mainly for improvements to the methods traditionally used to fight floods. The levees, revetments, and channel projects along the Boise, including the ones constructed by the Corps, had always been selectively located to defend the most vulnerable places. No one had ever systematically planned a substantial levee system to protect long stretches of flood-plain land. The cost of such a system would not have justified the benefits that might have been obtained, even though row crop prices had risen. Severe floods did not occur often enough.

Urban interests were missing from the policy discussion. The city of Boise had sustained such minor damage that it did not see its interests particularly affected by the threat of another flood. The city had prepared itself for flood dangers long before 1943. For the most part the city had either protected new development behind levees or filled the flood plain. Mayor Walker's list of itemized damage expenses had been remarkably short, hardly the stuff of disaster: McCaslin Lumber, $125 to raise materials off the yard and first floor of their warehouse; Boise Water Company, $500 to protect a pump station; and homeowners, $400 to move goods and clean out basements. Twenty other establishments spent $3,590 among them to move merchandise out of danger. Public costs for bulwarking the levees and repairing city streets and bridges amounted to $34,473.81.[8]

The heavy bridge damage itself did not inspire any proposals to change policy. All interests, urban and agricultural, were affected by bridge damage, a perennial hazard

Idaho State Historical Society #64-74.32
**Figure 2.3**
**Austin Walker, mayor of Boise from 1942 to 1945.**

*After the Flood*

**Figure 2.4**  *U.S. Army Corps of Engineers*
During an aerial flight in 1943, a reporter counted at least 150 sheds, barns, haystacks and outbuildings surrounded by water. Here the river challenges the Notus Bridge.

of flooding. But local government budgets tended toward the repair of failed bridges rather than their replacement with more costly ones better designed to withstand the known levels of flooding. It was cheaper to straighten the river channel upstream of bridge abutments and reduce the risk of future damage. The disruption of farm-to-market transport across the river had not reached the stage where the capital expense of flood-proofing bridges was a high priority for community funds. (It would not be until 1985 that Boise officially required that all bridges, both pedestrian and auto, must be constructed to withstand a 100-year flood.[9])

It was the Boise Valley reclamation interests who articulated and advanced the winning solution to Boise River flooding. Their idea would not only solve flood problems for farmers in the valley, but also promote more desert irrigation. Their point of view was well summarized by Watermaster William Welsh, who wrote to Governor C.A. Bottolfsen in April 1943 before the flood had yet receded:

> Personally, I believe that the proper way to control these floods whenever possible and especially here in the Boise River is by the construction of reservoirs. As the Army Engineer, Mr. Charles F. Beattie, who visited the Parma section with me yesterday said: "By the construction of reservoirs you will kill two birds with one stone." First, a reservoir controls the floods, and second, it stores the water and makes it available for irrigation.[10]

Although Mr. Beattie considered new reservoirs to be a practical solution, it remained for his Portland and Washington superiors to come to that conclusion.

## Federal Responses

Having been asked by Welsh and others to repair the damaged flood levees on the Boise before the flood season of 1944, the Corps was reluctant to do it. The engineers felt that spending resources upon the Boise River for emergency work was wasteful because the repairs would simply be submerged again should there be another flood. The Corps was, however, more inclined to "study the need for additional flood storage in the headwaters of the main stream, and to investigate thoroughly the desirability of insuring that the operation of the multiple-purpose reservoirs in the interest of flood control will not be subordinated to the interest of irrigation."[11] Idaho's congressional delegation, always in close touch with reclamation interests, cared little for the Corps' reluctance about emergency levees and prodded the Corps to consider the need for both elements. The Corps eventually launched a two-pronged program: to make emergency repairs on the levees against a potential flood next spring and to conduct a new study aimed, at least in the minds of the irrigation interests, at developing another reservoir.

The irrigation interests would find that their needs dovetailed nicely with the concerns of national leaders already planning for post-World War II America. National policymakers feared that declining production and employment after the war would cause the nation to retreat into economic depression. One safeguard against this was public works projects, for which federal agencies were directed to make plans that could be undertaken when the war was over. In the area of water projects, there were two federal agencies in a position to compete for authorization for similar projects—and each was eager to grow and expand its influence. One was the Bureau of Reclamation, which regarded the western states as its turf. The other was the Corps of Engineers, which regarded flood control as its turf everywhere in the country, including the West. Lucky Peak Dam was to become one of the pawns in the struggle between these two agencies.

The origins of this situation went back to the mid-1920s. While local communities like Boise were straightening river bends and filling lowlands, the idea of coordinated river basin development began to interest Congress. In 1927 Congress directed the Corps to analyze how multiple-purpose water projects in the Columbia Basin could achieve the greatest benefits "for the purposes of navigation and efficient development of its water power, the control of floods, and the needs of irrigation."[12] This study and others like it became known as the 308 Reports, after the number of the House Document that recommended the studies. The North Pacific Division of the Corps completed the Columbia Basin report and submitted it to Congress in two parts, one on the Columbia River and its tributaries, the other on the Snake River and its tributaries.*

The reports and their recommendations, printed in 1933 and 1934, concluded that "there are no important flood problems in either of these sections." For the Snake River and its tributaries, in fact, the Chief Engineer of the Corps recommended that improvements in the Snake River Basin for power, navigation, flood control, irrigation or "any combination thereof is not advisable at the present time" except for already authorized projects.[13]

While flood problems on the Columbia did not generate Corps enthusiasm, serious floods elsewhere in America stimulated legislation in 1936 that gave the Corps national responsibility for preventing flood damage.[14] Along with the 308 Reports, the multiple uses of water had to become part of the logic of any Corps planning involving flood control, particularly in the West. Meanwhile, the Bureau of Reclamation had long been involved in multiple-purpose projects that combined irrigation with electrical power generation because of the frequent need for pumping in order to deliver water.[15]

During and even before World War II it became apparent that the 308 Reports had underestimated the amount of power that would be needed in the Northwest after the construction of Bonneville and Grand Coulee Dams. In 1943 the Senate directed the Corps to review once more water resource development in the Pacific Northwest.[16] The war years diverted Corps energies to military objectives, but eventually the agency spent $5 million and 500 person-years on an eight-volume report issued in 1948. The Bureau of Reclamation had finished its own competing report on the Columbia four months earlier. Both agencies proposed building a high dam at Hells Canyon on the Snake River, a huge project that would have irrigation and flood-control benefits and generate a great deal of power.[17] This project was to become another (very large) pawn in the match between the Corps and the Bureau.

The six years between 1943 and 1949 were the same years during which the local irrigation interests in Boise were maneuvering their hopes for a new reservoir through the Corps and congressional processes of defining and justifying a project, getting it authorized, and finally obtaining an appropriation to pay for it. If they were to succeed, it would have to be against a national backdrop in which the Corps also had to succeed in building its first dam in the Snake River Basin, a territory that had been the exclusive turf of the Bureau for over 40 years.

---

*The North Pacific Division is one of 11 Corps of Engineers divisions across the United States. Each reports to Corps headquarters in Washington, D.C., while several district offices report to each division.

# Chapter Three
# Lobbying for Lucky Peak

## Sweepy-Weepy and the Mountain Home Desert

In October 1943, six months after the flood, Harry W. Morrison, the president of an organization called Southwest Idaho Water Conservation Project, Inc., (SWIWCP) and Idaho Governor C. A. Bottolfsen sent synchronized telegrams to the Idaho congressional delegation outlining what they wanted: Since the War Department was "not favorable to anything except long-range program . . . hope you can impress upon . . . Army Engineers necessity for having work done this year in anticipation of disastrous flood . . . next spring." The next day Senator D. Worth Clark wired back the good news that he had gotten a resolution through the Commerce Committee authorizing the Corps to do both.[1]

The Southwest Idaho Water Conservation Project, Inc., the force behind the telegrams, was a

**Figure 3.1**  *Karsch, Ottowa*
**Harry W. Morrison, president of Sweepy-Weepy.**

private promotional organization, the voice of growth-oriented businesses supporting reclamation as the key to economic progress in southwestern Idaho. By 1940, when Morrison and other business leaders created SWIWCP, Arrowrock and the other irrigation reservoirs in Idaho had demonstrated dramatically what stored water could do for the economy. Existing farms prospered and new ones sprang up from the desert. Agricultural growth meant more business for bankers, suppliers, builders, railroads, and food processors. In the Boise valley these growth-oriented businesses constituted a powerful element of the local reclamation interest.

The promotional association (whose awkward initials became known conversationally as "Sweepy-Weepy")[2] wanted to accelerate the development of water and land resources in southwestern Idaho. It believed that Idaho's growth rate had stagnated since 1930 because the state was not getting its rightful share of federal reclamation dollars compared to sister states.[3] It promoted two specific goals only: to get more water for existing developments that needed supplemental water in low runoff years and to put to "beneficial use" all available water in Southwest Idaho.[4] A SWIWCP pamphlet declared:

> Each spring, water worth millions of dollars rushes to the sea in an impotent flood. When this flood is harnessed The Valleys of Tomorrow will become The Valleys of Today, a constant . . . agricultural storehouse for the nation.[5]

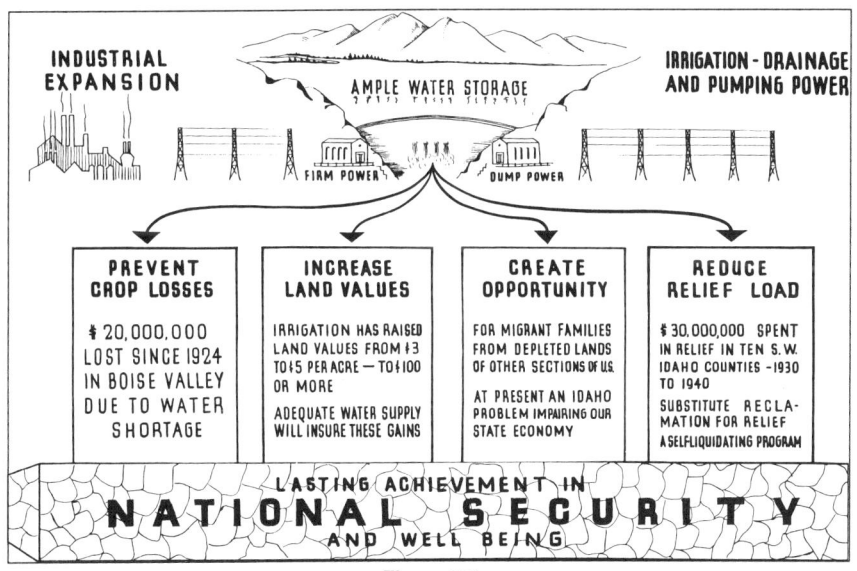

Figure 3.2
An illustration from SWIWCP's booklet *The Valleys of Tomorrow*.

SWIWCP determined to take a rational and energetic approach to attaining its goals. Its executive committee members realized that they would have to maintain continuous contact with relevant government agencies and bureaus in Washington, D.C. They would have to keep technical reports up-to-date, assiduously monitor legislation, keep their political strategies current, and nurture a strong communication link with agricultural interests through publicity and other methods.[6] The organization was well financed, highly organized, and always alert to opportunities that might advance its goals.

The SWIWCP leaders were among the most powerful and influential people in the state of Idaho—bankers, contractors, business and political leaders. Harry W. Morrison, the organization's president, was the co-founder of Boise-based Morrison-Knudsen Company, one of the builders of Hoover Dam and a score of other dams during the Depression. In 1940 the SWIWCP executive committee included Lynn Driscoll, president of First Security Bank; T.M. DeCoursey, Chairman of the Board of County Commissioners in Canyon County (the county covering the lower half of Boise Valley and adjacent to Boise's Ada County); C.C. Anderson, founder of the Golden Rule department store chain (later the Bon Marche); and Harry L. Yost, the postmaster. Later, Boise River Watermaster William Welsh joined, as did J.R. Simplot, founder of the J.R. Simplot Company and supplier of potatoes to the United States Army during World War II; and Robert Wetherell, an influential state senator from the town of Mountain Home in the 1950s and in the real estate business there.[7]

Financial support for the association came from chambers of commerce in Southwest Idaho, banks, railroads, utility companies, ditch companies, businesses, and corporations. As part of its communication and publicity function, SWIWCP structured itself as a mass membership organization and sought wide involvement from the agricultural community of southern Idaho, hoping for at least 10,000 members.[8] The association's conduit of information to and from federal offices in Washington, D.C., was lobbyist Everett W. Rising, a former secretary of a local chamber of commerce and vice president of the Western Beet Growers Association. Rising had been an agent for the Idaho Power Company in the 1920s and for Morrison-Knudsen in the 1930s. He had extensive connections with the National Reclamation Association, a major irrigation lobby, and still worked for Morrison in the 1940s and early 1950s.[9]

Throughout the early 1940s, SWIWCP had been working with the Bureau of Reclamation to develop an irrigation scheme for the Mountain Home Desert, an area of considerable potential southeast of Boise on the north side of the Snake River. Reclamation interests had long wanted to irrigate this desert, but had been thwarted by the problem of finding a source of water, discarding one unfeasible plan after another. They had first dreamed of irrigating this fertile part of the Snake River Plain as early as 1893 when they had considered pumping water up from the

Snake River Canyon. Later ideas for transporting water from some other basin, such as the Boise or even the Salmon River, had, like the earlier ones, never been affordable.[10]

In 1944 the Bureau of Reclamation and irrigation interests conceived a promising new idea. A Bureau report released that year showed how water from the Payette River could replace Boise River water presently irrigating Boise Project lands, freeing the Boise River water for pumping further south to Mountain Home.[11] The scheme required two storage reservoirs on the Payette River (Cascade Reservoir on the North Fork, and Garden Valley Reservoir on the Middle and South Forks), two power plants to pump water uphill over two divides, and a 50-mile canal to transport water from the Payette to the Boise drainage. The canal would course through tunnels, run parallel to the Payette River past Horseshoe Bend, and then wind its way through the foothills north of Boise to meet the Boise River at Diversion Dam. There another lift into a "Hillcrest Canal" would send the water to the lands of the Boise Project. Meanwhile, a diversion, lift station, and tunnel would be built just south of Anderson Ranch Dam to carry Boise River water to new lands west and south of the town of Mountain Home. (Figure 3.3).

This remarkably complex scheme would have cost $253,234,000 (in 1944 dollars) and placed 200,000 acres of sagebrush desert under the plow. The Bureau estimated that annual crop income would be $30 per acre and that the charge for water would be $4 per acre, a high cost considering the total of all other costs. Benefit/cost and other analyses concluded that farming never could repay its share of the project's cost—or even a substantial part of it.[12] Clearly, this was a major obstacle to the plan.

There were additional obstacles, such as the engineering of the Horseshoe Bend Canal. With 24 miles of the canal proposed to go through a tunnel, the initial construction cost would be very high. Because of porous soils, the water lost as seepage would be substantial. Worst of all, the terrain in the hills around Horseshoe Bend and above Boise is subject to landslides, slumps, subsidence, and rapid erosion. If the canal ever broke above Boise City, it would cause massive erosion and calamitous flooding, so annual maintenance costs for such a canal would be extraordinary.[13]

Despite its costs and hazards, the project was regarded as the only key to further growth in Southwest Idaho. SWIWCP leaders—and most of the political and other leaders in the state—linked Idaho's growth and progress almost exclusively to the expansion of irrigation, giving little thought to industrialization. In the West, capital for irrigation reservoirs came from the federal government. It was well understood that private capital could not undertake the quarter billion dollar investment to irrigate the Mountain Home Desert, the largest and one of the last unrealized irrigation projects remaining in the entire Columbia River system. With this rich soil under the plow Idaho could be the breadbasket for the booming

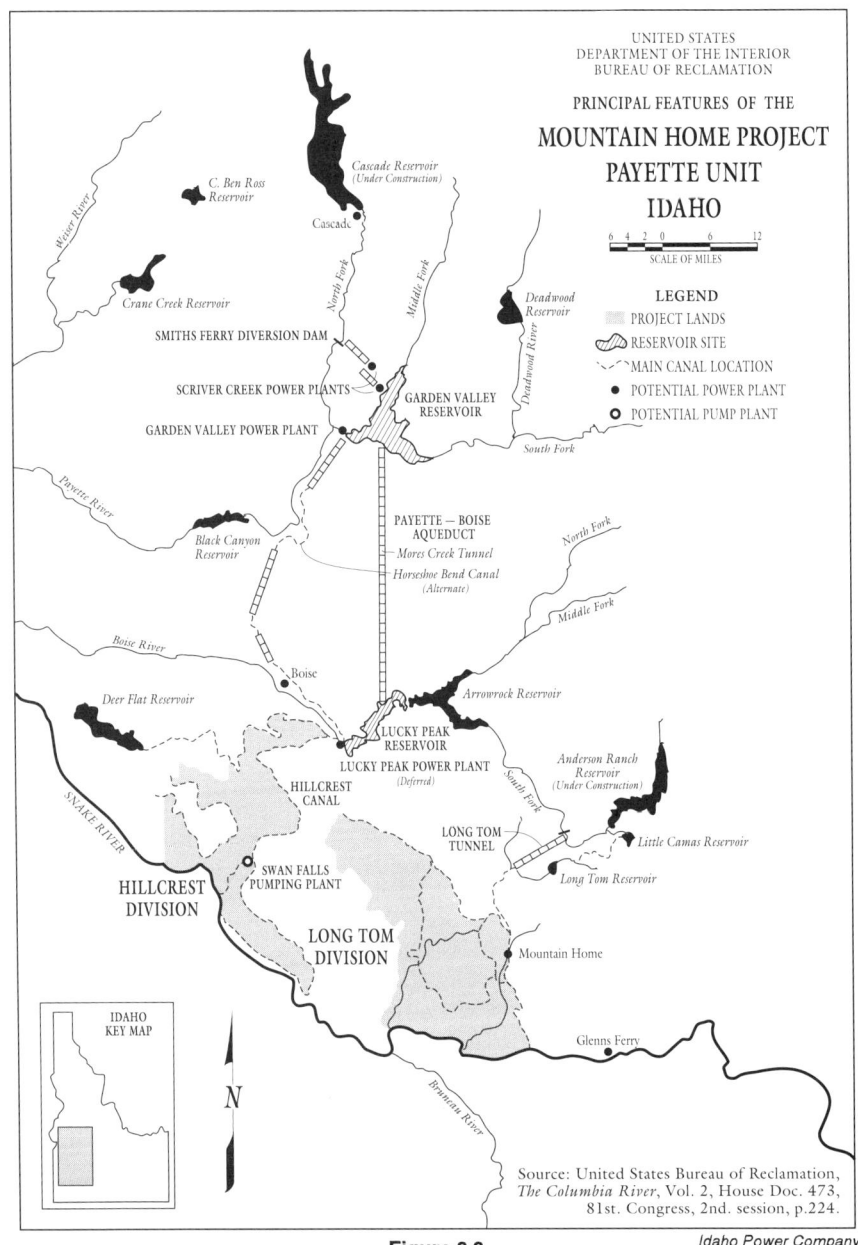

**Figure 3.3**
The Mountain Home Project would have transferred water from the Payette River Basin to the Boise River Basin, and water from the Boise River Basin to the Snake River Basin.

industrial populations of the Pacific Coast. The 1943 Boise River flood offered an opportunity for irrigationists to join forces with the flood-prone farmers, who wanted emergency protection against a potential flood in 1944, and the Corps of Engineers, who were more inclined to build another reservoir as a permanent solution to flooding. Such were the visions of those who lobbied for Lucky Peak Dam.[14]

## The Corps Plans Lucky Peak

With the blessings of D. Worth Clark's congressional subcommittee, the Corps immediately prepared plans and specifications for a new Boise River dam and reservoir. In December 1944 the engineers were ready to come to Boise and present them to the public. Watermaster William Welsh coordinated a series of meetings in Boise to explain how the project would promote irrigation and prevent flooding. He invited the farmers who had suffered in the 1943 flood to a hearing at the Statehouse and invited groups of community and political leaders to luncheon and dinner gatherings at the Hotel Boise.

The first order of business at the hearing was to present the Corps' proposal. Corps engineer Stanley Sporseen began by explaining that a coordinated levee project was not justified because flood control did not offer enough benefits. "We have studied levee protection to some extent, but the only benefit you get from levees is flood protection and that isn't enough to pay for the levees, so that practically has been thrown out," he said.[15] Flood-damage potential was not severe enough to justify a levee system. If that were the case, how could a reservoir, more costly than a levee system, be justified if flood control was the only benefit?

It could not. The beneficiaries of a purely flood-control project would have been the farmers who operated in the flood plain, the taxpayers whose funds paid to guard levees and maintain bridges, and the canal and ditch companies for whom maintenance costs might be reduced with less flooding and siltation. Despite the evacuations, emergency preparedness, agricultural damage, and loss of efficiency during the flood crisis, these damage potentials could not supply enough economic benefit for a favorable benefit/cost calculation—neither for levees nor a dam. But when desert reclamation interests threw irrigation benefits into the equation, the Corps managed to justify Lucky Peak Dam as a multiple-purpose project.

The Corps officer went on to describe the reservoir project. The dam would be built about a mile above Diversion Dam on a foundation provided by hard lava rock.[16] Besides the main Boise River flow, a dam there would catch the flow of Mores Creek, the one last tributary stream not controlled by either Arrowrock or Anderson Ranch dams. The dam would raise the water level in front of Arrowrock Dam by 95 feet and cover several of the dam's drainage ports.

The engineer listed the essentials: reservoir capacity of 306,000 acre-feet, 26,000 acre-feet of dead storage; a compacted earth-fill dam 251 feet high, 1,700 feet long, 24 feet wide. He presented details for the spillway, outlet capacity, and the diameter of the outlet works tunnel. The Corps named the project Lucky Peak after a land feature that happened to be on the same topographical map that showed the reservoir site.[17] Colonel Ralph Tudor, District Engineer for the Portland District, said the dam would be able to control the largest recorded flood on the river, that of 1896. Although he stressed that the plans were "preliminary," they were to change little as planning progressed to construction.[18]

The second part of the hearing gave the audience, mostly farmers and public officials who had been affected by the 1943 flood, a chance to tell their stories and express their hopes for federal help in solving their flooding problems. Only one individual had anything critical to say about the Corps' proposal. Mining engineer and former state senator R.E. Whitten got up to state that he disapproved of the poor engineering of local bridges and the inadequate maintenance of the river channel. He endorsed "legitimate, economical, and justifiable" flood control, he said, but this particular project smacked of "political pork," and cost too much money with the war debt accumulating. Besides, Arrowrock and Anderson Ranch would take

**Figure 3.4** *Idaho State Historical Society #69-148.9*
**Aerial view of Lucky Peak Dam site in 1950, looking upstream.**

*Lobbying for Lucky Peak* 31

care of the flood problem.[19] Whitten's comments are noteworthy partly because they were somewhat unique. Further serious opposition to Lucky Peak Dam never materialized.

After the hearing, the Corps entourage met with business and political leaders in the dining rooms of the Hotel Boise. Here engineers talked more about the irrigation benefits of the project, benefits that SWIWCP leaders already understood. Lucky Peak would make 150,000 acre-feet of water available for irrigation. This water would make it possible to use water from Anderson Ranch Reservoir to irrigate the desert area south of Boise near the town of Mountain Home. A power plant at Arrowrock could generate electricity primarily during the irrigation season with Lucky Peak acting as a re-regulating reservoir. To make this complex series of benefits possible, all three reservoirs—Arrowrock, the unfinished Anderson Ranch, and Lucky Peak—would have to be operated together as one system. Only by coordinating all the operations could the Corps maximize Lucky Peak's potential for both irrigation and flood control. Since a workable plan would have to rely on the most accurate water flow predictions possible, new hydrological stations in the mountains would be established to improve the forecasting of snow and water conditions.[20]

**Figure 3.5** *Idaho State Historical Society #2647*
**Anderson Ranch Dam, completed by the Bureau of Reclamation in 1950.**

The reclamation interests were eager to help the Corps devise a workable plan for Lucky Peak. Project promoters had seen at once that Lucky Peak Reservoir presented an opportunity for a shorter canal route between the Payette River reservoirs and the Boise River. If Lucky Peak's lake were to serve as a receptacle for the water, the canal could be 20 miles shorter and avoid the dangerous foothills. The tunnel would be just as expensive, but the more direct route would reduce other risks and some of the maintenance costs. The water could be dumped into the reservoir and held there until needed instead of being pumped immediately out of the Boise River canyon.

The pre-Lucky Peak plan showed water being pumped from the level of the Boise River in order to get it over the hill. Since the Lucky Peak reservoir surface would be higher than the river, the pumps would have less of a lift, and less electrical power would be required. This reduced power lift was to become one of the non-flood benefits of Lucky Peak.

SWIWCP geared up to promote Lucky Peak as part of the Mountain Home Project. In 1945 Rising reminded Morrison that federal funding formulas, which required that farmers reimburse part of the construction costs, would not cover the Mountain Home Project. It was $77 million too expensive. There was no hope for getting Congress to authorize all of the many components of the plan as one major project.[21] SWIWCP settled on a strategy to get portions or segments of it authorized independently—hoping in this way eventually to accomplish the goal.[22] Later the association would promote the construction of a high dam at Hells Canyon (on the stretch of the Snake River that forms a boundary between Idaho and Oregon) with the hope that its power sales could repay the costs of the Mountain Home Project.

The Corps of Engineers was aware that Lucky Peak was useful to the Idaho reclamation interests. At the time, the federal government financed flood-control projects without requiring cost-sharing or reimbursement by local interests. For multiple-purpose projects, only the portion of the cost not attributed to flood control had to be reimbursed (albeit on favorable terms). Even though Lucky Peak was but a small part of the Mountain Home Project, it was one step on the path to a complete irrigation project.

## SWIWCP and The Corps Justify Lucky Peak

Reclamation interests were useful to the Corps as well. Major General Thomas M. Robins, Acting Chief of Engineers, wrote to Congressman Henry Dworshak that perhaps the Corps could operate this new reservoir without irrigators claiming the water first.[23] Nevertheless, planning involved close work with the Bureau of Reclamation and the existing water users. The Corps had to develop a concept for coordinating the operation of the three dams without creating anxiety about the prerogatives of the Bureau or the rights of valley water users. The two agencies agreed that the Corps would build the dam and then turn it over to the Bureau to operate.

Early in 1946 the Corps finished its Boise River survey and recommended to Congress that Lucky Peak Dam be built as a multiple-purpose project to benefit irrigation and flood protection.[24] The law required that private water users reimburse the government for their share of the project cost; however, the Corps analysis concluded that reimbursements should either not be paid or should be postponed until the irrigation and power components of the Mountain Home Project were built. The first step of the analysis demonstrated that the benefits of the project outweighed the cost. This procedure involved annualizing the costs over a project life of 50 years using an interest rate of 3 percent. The annual benefits had to equal or exceed the costs. Then the various beneficiaries were allocated the percentage of the costs they should pay.

As the Corps had been reporting since its public hearing in 1944, the annual cost for flood protection alone exceeded the benefits. Now it identified the exact amount of this excess annual cost as $62,820.[25] Also as predicted, irrigation benefits made up the difference. These benefits were of two types, and they neatly paralleled the goals of SWIWCP. The first was that Lucky Peak Reservoir would provide supplemental water that would help end shortages in years of low runoff, SWIWCP's first goal. Lucky Peak would hold 280,000 acre-feet of active storage; when combined with the other three reservoirs, this meant that 1.213 million acre-feet, or nearly two-thirds of the Boise's average yearly runoff, could be conserved for irrigation.[26] The Corps had some difficulty estimating the economic value of supplemental water because such water had only been critical during one year in the previous 20 years, with minor shortages in two others. Supplemental water had no value in the years when it was not needed. The Corps annualized the pre-Lucky Peak shortage at 11,500 acre-feet and assigned to it a nominal value of $1 per acre-foot, a total annual benefit of $11,500.[27]

The second irrigation benefit was the reduction of the pumping lift from Lucky Peak Reservoir as outlined by the proposed, but not authorized, Mountain Home Project. The Bureau of Reclamation had calculated for the Corps the advantages for the Mountain Home Project of building Lucky Peak with a permanent pool (level at which water would not be drawn down) 80 feet higher than the river surface. Instead of lifting 285 feet over the divide, the pumps would work to lift only 205 feet, for an annual power savings of $85,400.[28] The calculation also elevated the benefits of Lucky Peak up over its cost, making the project legally feasible for the Corps and contributing to SWIWCP's second target, the use of water for additional irrigation development.

If a power plant were built at Arrowrock Dam as part of the project, another $89,200 in power sale benefits could be added. Although the original design for Arrowrock had provided for future hydroelectric power, installation had not been practical because discharging stored water was often incompatible with irrigation. With Lucky Peak's reservoir downstream, the water used for power generation could be stored there (re-regulated) and saved for later irrigation use.[29]

The Corps mentioned other small or non-valuated benefits. Since it would be building an access road to the reservoir and a park nearby, it assumed that the recreational benefits would at least equal the cost of $2,000, a valuation which seems underrated from the perspective of the 1990s.[30] Other benefits were mentioned, but not given a dollar value, perhaps because they were public goods and too difficult to value in an economic sense. These included enhancements to fish and wildlife that would accrue through "better regulation of stream flow," prevention of loss of life,[31] allaying fear of floods, expansion of local business and residential areas, enlargement of the local tax base, and increased security of all the residents of the valley. In any case the Corps had made its mathematical case and needed no additional figures for the "benefits" side of the equation.

After adding up the dollar totals, the benefits outweighed the costs by 1.25 to 1, a ratio potentially acceptable to Congress since the law required that flood control projects have at least a one-to-one ratio.[32] Now the second stage of the analysis could proceed: determining who would pay for what, the delicate matter of allocating costs to the beneficiaries. The Corps methodology allocated the construction costs based on the percentage of the benefits a private interest received. Flood-control benefits were 70 percent; since these would benefit the general public, they would be built at federal expense. This would suggest that private parties benefiting from irrigation and power benefits would have to reimburse the government for the remaining 30 percent. But each of the allocations, as explained below, was analyzed away, and Lucky Peak became a 100 percent federally funded project.

The allocation of costs went as follows:[33]

|  | Construction Costs | | Annual Costs | |
| --- | --- | --- | --- | --- |
|  | Benefit % | $ Cost | Interest | O&M |
| Flood control | 70.3 | 8,340,000 | 332,200 | 25,000 |
| Irrigation | | | | |
|   Supplemental water | 1.7 | 202,000 | 8,000 | 0 |
|   Reduction of lift | 13.5 | 1,601,000 | 63,000 | 0 |
| Power | 14.1 | 1,672,000 | 66,630 | 10,000 |
| Recreation | .4 | 50,000 | 2,000 | 0 |
| TOTAL | 100.0 | 11,865,000 | 471,830 | 35,000 |

The Corps decided that the irrigation operations at the dam (which consisted primarily of passing water through the system from Arrowrock Reservoir) would contribute such a negligible amount to O&M (operation and maintenance) cost that they need not be allocated to irrigation.

*Lobbying for Lucky Peak*                                                                                     35

Then the Corps estimated that "irrigators in the Boise River Valley could afford to pay approximately $1.00 per acre foot for supplemental water supplied by transbasin diversion from Payette River." On this basis, repayments for supplemental water would be less than 2 percent of project costs and be difficult to trace to a precise user. The Corps decided to forego collecting it, pointing out that those users were the same ones still paying for Arrowrock Dam, and since the Corps was not going to repay them in any way for the use of Arrowrock in the new three-reservoir management system for flood control, the additional water from Lucky Peak could be considered "recompense."[34] In other words, for cooperating with the Corps flood-control project, the water users would be getting more water and not paying for the reservoir that made it possible or for the cost of maintaining it.

Everyone agreed that the future Mountain Home irrigators would not be able to pay the cost of pumping water to their land. Therefore, the Corps reasoned, the power plant at Arrowrock would not be saving the water users any money. The benefit would instead be defined as the surplus power available for sale because the pumping lift was not as great as originally planned. Since the future power market was very difficult to predict, the Corps proposed that the matter be postponed until pumping to the desert had actually begun. Then the Corps could deal with the Bureau of Reclamation via a power sale agreement, and an allocation of costs to power could be determined.[35] It was the irrigation benefits, however, that had contributed so much to the positive benefit/cost analysis.

The Corps acting division chief of San Francisco, Colonel S.E. Nortner, who reviewed the Corps' report, made it clear that the power and pumping lift benefits would not count right away. Without them, the benefit/cost equation was $458,310 to $460,000, only $1,290 shy of a one-to-one ratio, but these numbers did not account for the intangible benefits and the irrigation-related options for the future. He declared, therefore, that the plan "is fully justified."[36]

Even though Anderson Ranch had been authorized just a few years before with flood-control benefits, the standard for flood control on the Boise had gone up. With only Anderson Ranch, the system would have controlled all but 21 of the floods that had occurred since 1865, but with Lucky Peak, it would have controlled all but two of them—that is, held them to less than 10,000 cfs.[37] The Boise would indeed be one of the best flood-controlled streams in the United States.

## Congress Approves Lucky Peak

During the spring and summer of 1946, Idaho interests labored to have the Lucky Peak project included in an omnibus flood-control bill. The *Review of Survey Report* was not approved by the Corps' Board of Engineers for Rivers and Harbors until May 1, after House hearings in April on the original bill. The Senate had to add it as an amendment.[38] The Idaho delegation communicated progress on the bill to William Welsh, who in his own letters and memoranda signed himself as Boise River Watermaster, or as SWIWCP member, or as Secretary of the Idaho State Reclamation Association, whichever was most appropriate to the matter at hand. His press releases on the progress of the bill typically mentioned both flood-control and irrigation benefits of Lucky Peak.[39]

When the Senate Commerce Committee heard the Lucky Peak amendment, the project ran into some trouble. The understanding between the Corps and the Bureau that the Corps build the dam and then turn it over to the Bureau did not sit well with the committee. Chairman John Holmes Overton (D-Louisiana) pointed out that the Bureau had built or planned 16 other dams in the region and that this would be the first one by the Corps. Committee members were unconvinced that benefits exceeded costs, and they asked that the benefits be broken down for agricultural and rural areas. Bureau of Reclamation Commissioner Michael W. Strauss, perhaps unhappy with the quiet arrangements reached in the field or concerned about potential damage to the Bureau's Arrowrock Dam, pressed for a congressional commitment that the Bureau would operate the dam.[40]

But when the discussions had ended, the committee resolved to approve Lucky Peak and directed the Corps of Engineers to operate it. To assuage irrigators' concerns that the reservoir would back up to the face of Arrowrock, the committee ordered that Lucky Peak should be "so constructed as not substantially to damage the structure of the Arrowrock Dam and shall be operated in such manner as not materially to interfere with the operation of said Arrowrock Reservoir."[41] After that, the bill progressed without further trouble, and President Harry S. Truman signed the Flood Control Act into law on July 24, 1946. Its amendment authorizing Lucky Peak, the third big reservoir on the Boise River, survived intact, a tiny part of the bill authorizing flood projects valued at $900 million all over the country.[42]

The Corps continued the field work, hydrologic analysis, engineering investigations, surveys, and all of the other complex planning required to design a dam. It was time to determine how much the project would cost, how much Congress needed to appropriate, and how many years it would take to build. Together with the Bureau, the Corps worked out the details of a river management plan to improve performance for both irrigation and flood control.

The next objective was to get Congress to appropriate funds to build the project. The Corps updated the costs in *Definite Project Report on Lucky Peak Dam, Boise River, Idaho* in October 1949.[43] The 1946 *Review of Survey Report* had estimated costs on the basis of 1940 values. Nearly 10 years later, costs—and a more realistic appraisal of the work to be done—had risen substantially. For example, the Corps had estimated earlier that all needed road and utility relocations would cost only $1.891 million. But in 1949, just relocating Idaho State Highway 21 and building a new bridge across Mores Creek would cost $6.89 million. The project total doubled to over $22 million. The annual cost rose to $962,000.

Having upgraded the costs, the Corps also had to upgrade benefits; this time the Corps projected more value for flood-plain development due to its removal from flood threat. The value of prevented damages went up to $867,000 each year, but this amount was still not enough to justify the project.[44] The other benefits were duly added for supplemental irrigation, recreation, and a new one for reduction in downstream siltation. Power benefits continued to be deferred. The new list of benefits gave a new total value of $1.069 million, an amount that exceeded the costs by enough to support the appropriation.[45]

Because the Mountain Home Project was still no closer to authorization than it ever had been, the *Definite Project Report* proposed an "initial plan" and an "ultimate plan." The initial plan would use all the storage space in all three Boise River reservoirs for combined maximum benefits of flood control, irrigation, power (at Anderson Ranch), and recreation. In all, the plan set allocation for dead storage, permanent pool and silt control for Anderson Ranch and Lucky Peak at 101,000 acre-feet. The remaining 983,000 acre-feet were to be operated for the other interests. The "ultimate" plan was for the day, thought to be within 15 years, when water would be coming into the reservoir from the Payette River, and the Boise's water would go the desert.[46]

By the time the appropriation bill was on the congressional table, the Corps and the Bureau had been working together on another "arrangement." Their competition to build a high dam at Hells Canyon was about to end. In 1948 the Columbia River flooded the town of Vanport near Portland, causing serious loss of life and property. The tragedy inspired President Truman to order the Corps and Bureau to coordinate and harmonize the recommendations in their separate Columbia River Basin planning reports.

The regional offices of the two agencies, alarmed at the possible consequences of a proposal to enact a Columbia Valley Authority similar to the Tennessee Valley Authority, collaborated with a will. Officers in charge of creating the terms of agreement were Robert J. Newell, regional director of the Bureau, and Col. Theron D. Weaver, North Pacific Division engineer. These two men "were always on the best of terms and neither inter-agency rivalries nor Washington directives disrupted the per-

*U.S. Army Corps of Engineers*
**Figure 3.6**
**Brigadier General Theron D. Weaver.**

sonal rapport which existed between the two," according to a North Pacific Division history. The Newell-Weaver agreement submitted in February 1949, however, did not entirely please their Washington superiors (the historian did not say why), who agreed to a different settlement on April 11, 1949.[47]

The agreement divided up the Columbia Basin into what one observer referred to as "spheres of influence" between the Corps and the Bureau. First, the Corps would retain the rights to build any projects that Congress had already authorized, including Lucky Peak. The Corps would give up its own ambitions to build the high dam at Hells Canyon and defer to the Bureau. Second, in dividing up future projects, the Corps would build all navigation and flood-control projects, while the Bureau could have all irrigation, drainage, and domestic water projects. That took care of everything except multiple-purpose projects—that is, projects that could reasonably be built by either agency. Here the two agencies simply divided the turf: the main stem Columbia below Grand Coulee would go to the Corps, and the Bureau would take the middle and upper Snake and most of its tributaries.[48] The consequence of the agreement was that the Corps of Engineers built and operated Lucky Peak, the Corps' only multiple-purpose project within the "sphere of influence" otherwise apportioned to the Bureau of Reclamation.

Meanwhile, SWIWCP lobbied for Lucky Peak appropriations in an atmosphere of widespread regional agitation over a congressional bill to create a Columbia Valley Authority (CVA). The unpopular CVA proposal occupied center stage in public debates at the time, while Lucky Peak progressed with very little comment. Nevertheless, Everett Rising, SWIWCP's lobbyist, attended carefully to all the preliminary details leading Lucky Peak toward reality—getting the planning funds in 1948 and the first construction funds in 1949. He engineered strategy sessions with the Idaho delegation and coached Governor C.A. Robins and William Welsh when they appeared before a House or Senate subcommittee. Suitable testimonials were duly prepared and offered into the hearing records. The rhetoric escalated along with the costs of the project. Governor Robins sent a letter for the record to Idaho Representative Abe McGregor Goff that said, "I do not need to review for you the disastrous results of the floods along the Boise River. To delay Lucky Peak is merely to invite the devastation of another disastrous flood, leaving then but a depleted area to protect."[49] Such communications, including that one, usually added a reminder about the benefits for the pending Mountain Home Project.

As the Corps proceeded to design the dam and an operational system for the Boise River, its engineers accounted for the various potential power and irrigation developments that were anticipated for the near future. Working with the Bureau, they determined the capacities, levels, schedules, and release formulas for the three dams. They wanted an effective level of flood control without compromising irrigation rights. A last-minute complication arose when the Bureau suddenly decided that the plans should show power facilities at the discharge outlet from the

Garden Valley tunnel. Rising smoothed it over and, finally, Congress began appropriating funds. Rising hoped for a full appropriation of $3.5 million for fiscal year 1950, but because of the Korean War, funds did not flow as rapidly as either the Corps or SWIWCP would have liked.[50] In the summer of 1949 the Corps was ready to accept bids for the first construction phase. The main embankment contracts went to Morrison-Knudsen, Macco-Puget Sound, and Terteling.[51]

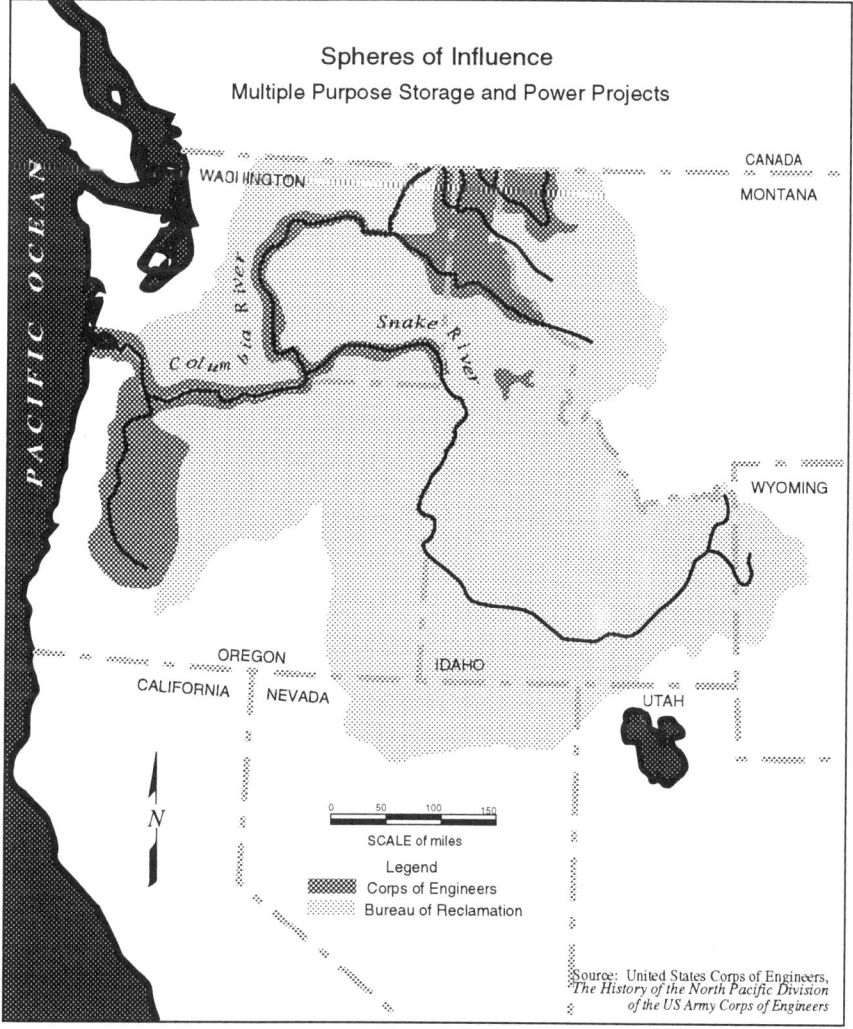

Figure 3.7
**The Corps and Bureau settled on "spheres of influence" for multiple-purpose projects in the Columbia Basin.**

## The Corps Builds Lucky Peak Dam

On November 30, 1949, over 400 people attended a ground-breaking ceremony hosted by the Boise Chamber of Commerce Reclamation Committee.[52] Lynn Driscoll, the master of ceremonies and chairman of the SWIWCP executive committee, watched Governor C.A. Robins and Mayor Potter Howard of Boise set off the first powder charge. A few days later he received a letter from Everett Rising, discussing the next moves in the campaign for the Mountain Home Desert. Rising outlined the many difficulties that SWIWCP still had to face, but in reviewing its progress to date, he expressed his satisfaction that Lucky Peak Reservoir "is now a definite unit of the Mountain Home Project."[53]

The Lucky Peak project had secured for SWIWCP its goal of supplemental irrigation resources in the Boise Valley. Further, it was an important contribution to the group's ultimate goal—agricultural expansion in Southwest Idaho.[54] "We know, of course," one of the SWIWCP associates had written in 1947, "that . . . the irrigation features of any multipurpose project will continue to enjoy a great advantage by having substantial portions of the costs allocated to flood control, wild life, and electric power. But that is the way it was always intended to be, that is the way it used to be, and that is the way we want it now."[55] With Lucky Peak, they had succeeded.

Because of the Korean War, the project still needed the attention of the Lucky Peak lobby. To offset wartime military expenditures, President Truman requested all departments not involved with the war effort to review their programs for ways of saving labor, money, and material. The Bureau of the Budget reduced the Corps' civil works budget by $550 million, and some projects had to be canceled or stopped. An anxious Everett Rising wrote to Harry Morrison, still president of SWIWCP, that if those interested would "write letters to their Congressmen giving reasons why the project should continue and the Congressional representative will, in turn, put up a fight for the project with the Corps of Engineers, their project may be kept going and something else closed down."[56]

The Corps District Engineer and his headquarters superiors agreed that Lucky Peak was a high priority. Expenditures for the project continued regularly until the dam was ready for dedication in the early summer of 1955. With construction underway, SWIWCP turned its attention to the next opportunity to promote the rest of the complicated Mountain Home Project. Financing was the major problem, but the proposed high dam at Hells Canyon presented a possible solution. Harry Morrison and other irrigation interests began advocating that revenues from the sale of public power at Hells Canyon pay for the Mountain Home Project.[57]

At the dedication ceremony on June 23, 1955, Harry Morrison sat on the guest of honor platform and listened, along with hundreds of people in the audience, to Undersecretary of the Army George H. Roderick say that "the objective of the

National Archives III-P-33299
**Figure 3.8**
**Army Undersecretary George H. Roderick.**

present administration [Eisenhower's] is to develop the full potentiality of all the great rivers of the nation." He called Lucky Peak a "monument to the might of the people" and the skill of the Corps, and reminded everyone that the dam would save a million dollars in flood damage every year. He went on to say that during the Korean War, while the country "is in desperate peril from Communist aggression," 85 percent of its resources were being spent on defense. Roderick repeated President Eisenhower's solution to the shortage of federal resources for river basin development—the "partnership concept" in building multiple-purpose dams.[58] Private and non-federal resources would augment or replace those of the federal government.

This policy of the new Republican administration had the ultimate impact of defeating the hopes of Morrison and others for the Mountain Home Project. The foes of the high Hells Canyon dam characterized it as a symbol of "creeping socialism" and supported the campaign of the Idaho Power Company to build three low-head dams in its place. The political battle engaged national attention, and the power company won the war. One of the fatalities was the Mountain Home Project. Later schemes to build and finance a water-delivery system all failed.[59]

Considering the furious opposition to the Hells Canyon project by advocates of "private enterprise," it is of interest to contrast the absence of similar protest to Lucky Peak. After mining engineer Whitten had characterized the project as "political pork" back in 1944, the only other criticism had come from newspaper editors, who objected to the high cost of the dam and professed bafflement as to why either the city or the county needed such expensive flood protection. When the Corps had first announced in 1948 that the cost of the project had doubled, the editor of *Statewide*, a Boise weekly, asked, "Has there been, or will there ever be enough damage to Boise Valley farms to warrant the expenditure of $21.66 million? Wouldn't that much money buy nearly all the lowland farms which might be flooded in the future? Would it stop a real rip-snorter anyway?"[60]

Later, when the dam builders celebrated the first water flowing through the diversion tunnel in 1952 with bus tours and other ceremonials, there was more editorial grousing. After a look at the Mores Creek Bridge under construction,

*Statesman* editor James Brown called it a "million dollar bridge to nowhere" and accused the Corps of giving the federal government an early cost estimate for the job, only to "discover" later that a few loose ends running into the millions of dollars needed to be cleaned up.[61] A year later, he suggested that as a flood-control project, Lucky Peak was a hoax.[62] "Everything seems in order but the possibility of a flood," he said. The city had never, he argued, needed such a gargantuan expenditure of federal funds.[63] But editorial irritations from *Statewide* and the *Statesman* were decidedly futile in retarding the progress of Lucky Peak.

The dramatic climax of the dedication ceremony finally came. Lucky Peak, born of a flood and ushered into being by the Idaho irrigation lobby whose principal interest was growth far distant from the flood plain, was about to begin its work. A turn of a wheel released 30,000 cfs from all six of the dam's manifold gates. Flip buckets directed the water up and out in a splendid energy-dissipating "rooster tail" spray, pleasing the crowd immensely.[64] But editor James Brown, still grousing, quoted a bystander who said, "the wonder is that it was ever built."[65]

*Idaho State Historical Society #69-148.17*

**Figure 3.9**
**The "million dollar bridge to nowhere," Mores Creek Bridge under construction in June 1953. The road below was inundated by Lucky Peak Reservoir.**

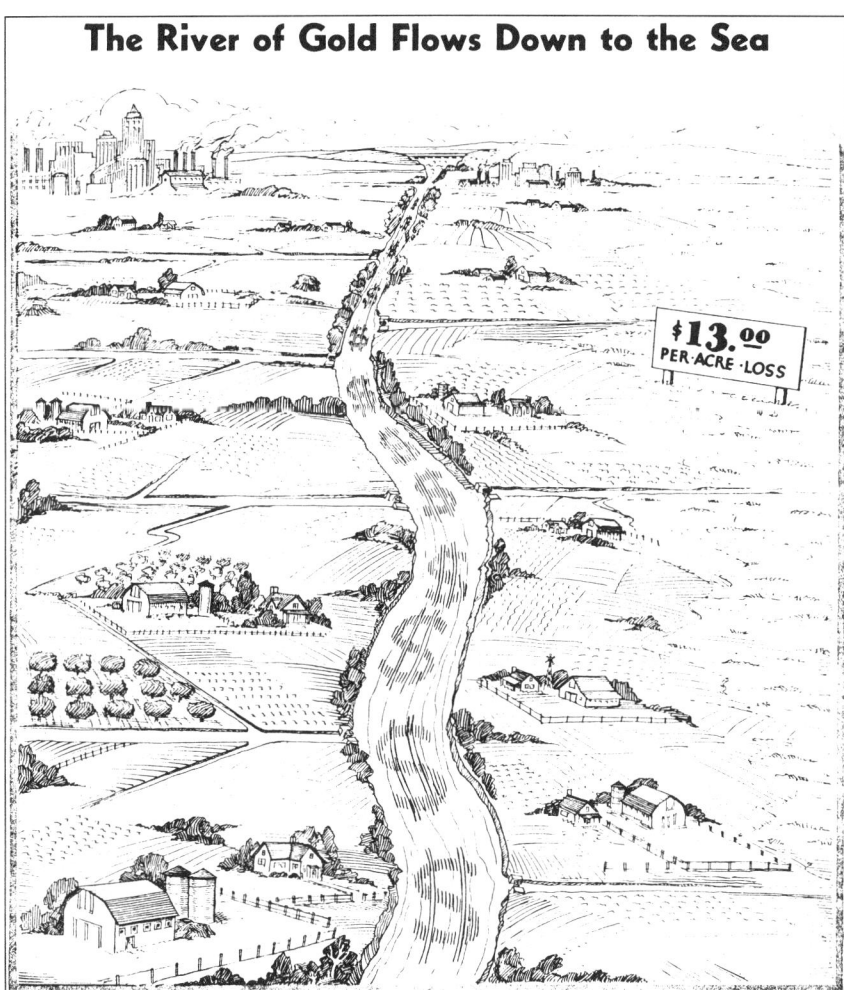

Figure 3.10

This *Valleys of Tomorrow* drawing expresses Sweepy-Weepy's attitude toward water in the 1940s and 1950s.

# Chapter Four
# Conflict Over Boise River Water

The SWIWCP organization faded from the Boise River scene along with the hopes for a high Hells Canyon Dam, but Lucky Peak, its legacy, remained. And the Boise Valley irrigators remained. The reservoir provided welcome security for low-water years by storing supplemental water. Lucky Peak made it possible to hope for genuine water security, not just for the current year, but for the next as well. Irrigators soon grew to resent the release of water for flood control—water they considered wasted.[1] With the weight of tradition and law behind them, the water users became a conservative force in the management of Lucky Peak and the Boise River's water, resisting any changes that might threaten their water supply.

Changes did threaten the supply. Lucky Peak was a multipurpose project, not solely for irrigation. It had the burden of flood control too. Conflict over Boise River water was probably inevitable because the inescapable logic of "multipurpose" was that no single use could be maximized at all times. The Corps described its management objective as "the optimum feasible joint use of the storage space in the three reservoirs for irrigation, flood control, power, and recreation."[2] The idea of joint use was hard for many irrigators to get used to, and they had to start getting used to it even before the dam was finished, because several interests different than their own emerged and began to demand water. Three distinct interest groups developed, each desiring to use Boise River water for a purpose other than irrigation. First, the fish and game interests asserted their desire to protect the Boise River trout fishery below the dam. Their demand for a minimum flow in the river ran up against the fact that Lucky Peak was not engineered with the needs of fish in mind. Then the flood-plain farmers noticed that spring flows were still damaging their lands. Their interests diverged from those of the desert irrigators and they started insisting on better flood protection. Finally, Boise developed both water-quality and taxpayer-protection reasons for needing a minimum river flow.

The common element among these three groups was their desire to have the Corps of Engineers send more water down the river instead of holding it in the reservoir—a complete reversal of the save-it-for-irrigation attitudes that had been dominant for decades. Since these competing demands for water directly affected the way the Corps of Engineers regulated the river, the Corps became both the target for pressure and a perceived instrument for solutions, even though it had several constraints on its range of action.

## Constraints on the Corps

It was impossible for the Corps of Engineers to satisfy all of the new demands for water, even if they had not conflicted with one another. As it began to operate the reservoir in 1955, the Corps had three constraints that shaped its response to the pressures of competing interests. The first was that the Corps built the dam with only one outlet tunnel for water, the second was the operating agreement the Corps signed with the Bureau of Reclamation in 1953, and the third was that Mother Nature is not perfectly predictable. Although the Corps forecasters did their best to predict the vagaries of the flood season each year, they admitted to anxious moments hoping their predictions of snow volumes and flood potentials would actually match the real events.

The first constraint was the design of the dam. On June 8, 1949, Lucky Peak engineers met in Portland with a board of consultants to review and approve the design concept for the dam. Among other things, they discussed how much seepage the dam would permit. All dams leak, although engineers like to use the phrase "water migrating through the dam." The design objective is to control this passage of water so it neither damages the structure nor reaches excessive amounts. The Lucky Peak engineers compared the merits of a more perfect seal (with a $500,000 price tag) to a less perfect seal that would allow a loss of as much as 30 cfs. After discussing the value of 30 cfs in arid country, the relative cost of the "positive cutoff," and the desirability of perfect confidence in the dam, the engineers decided on the more certain, and more expensive, cutoff method.[3] They also thought that the sight of water draining visibly from the dam would have an undesirable psychological effect on the public. This discussion appears to be as close as the engineers came to considering any passage for water other than the one through the main outlet tunnel.

That one outlet tunnel was 23 feet in diameter. The operators controlled the water release through a series of valves and conduits that directed the flow to six curved surfaces called flip buckets, which shot the water into the air to dissipate its energy and prevent erosion of the stilling basin below the outlet.[4] Whenever the tunnel and control works had to be inspected for maintenance or repair, there was no other route for water, so the operators just shut the river off as though the dam

**Figure 4.1** *Idaho State Historical Society #69-148.11*
**The first 40-foot section of the outlet tunnel liner on its way to the site in September 1950.**

were a spigot. They did this during the late fall or early winter months after the irrigation season was over. The only water entering the river below the dam dribbled in from downstream drainage ditches and groundwater seepage. Such low or no-flow conditions stranded fish and their spawn, killing them. But there was no choice—the dam had to be maintained and repaired.

The second Corps constraint pertained to the operating agreements that had been made with the Bureau of Reclamation. While Morrison-Knudsen and its partners still were erecting the dam, the Corps and Bureau got together to work out a plan of operation. They had to transform the three independently authorized dams into one well-coordinated system improving both flood control and irrigation. Although the Corps received recommendations from the fishery interests, public health agencies, and recreation advocates about how the system should function, only the water users sat at the negotiating table. The other interests' representatives were not invited to the negotiations, and the agreement did not deal with their concerns. In those days the Corps did not have the sensitivities—or the legal mandates—that would have encouraged broader participation in the decision-making. The "whereas" phrases in the agreement only listed those rights and prerogatives recognized by law, either by congressional authorizations or by Idaho's water rights statutes.[5]

After the Corps and Bureau regional offices had reached an understanding with the water users, they sent their work to the upper levels of each agency. On November 20, 1953, the Secretary of the Army and the Assistant Secretary of the Interior signed a memorandum of agreement, a succinct statement of the operating principles for the three dams that, although not an actual law, certainly carried the force of one.[6] The Corps and Bureau then prepared a *Reservoir Regulation Manual* to describe for operators the detailed routines for carrying out the agreement.

The agreement divided the year into two parts and set forth the operating procedures for each. The period between January 1 and July 31 was the flood-control phase; during this period, the Corps of Engineers was responsible (after consultation and agreement with the Bureau) for making flow-management decisions. For the rest of the year, the Bureau of Reclamation made the decisions. The Corps' flood-control phase itself had two parts: the "evacuation period" and the "fill period." During the evacuation period, the operators had to make sure there was enough storage space in the reservoir to catch whatever flood had been forecasted for that year. This meant releasing water to flow down the river. The goal was to release no more than 6,500 cfs, which was the amount considered to be the river's bankfull capacity. It would turn out that in years when the forecast was for extremely high runoff, they would have to exceed 6,500 cfs in order to empty enough of the reservoir in time to have room for the flood. The worst thing that could ever happen would be to have a full reservoir and lose control of the flood altogether—as Arrowrock had lost control of the flood in 1943. To the irrigators, water released to "make the space" for the flood looked like wasted water, since it was going down the Boise, down the Snake, and out to the Pacific, since there were no irrigation diversions in those winter months of the evacuation period. If all went well, the fill period began March 1, about the time natural mountain runoff began to exceed 6,500 cfs.* When the danger of a spring flood had passed, the stored water was available for irrigation and other uses later in the year.

An ideal operation thus depended upon reliable water runoff forecasts and a minimum of surprises from Mother Nature. If forecasters underestimated the amount of moisture in the watershed and failed to make room in the reservoirs early enough, they risked a flood, and the release rate would have to exceed 6,500 cfs. In the worst case, the reservoirs—Lucky Peak and the two above it—would all fill too soon and the river below would have to take whatever was running off the mountains. Unlike Arrowrock Dam's limited outlet capacity of 10,000 cfs, Lucky Peak could release over 30,000 cfs through its outlet tunnel. The dam designers

---

*Because some of the water released from the dam would be diverted to Lake Lowell via the New York Canal, river flows were measured below Diversion Dam. All references to flood flows hereafter refer to measurements taken in Boise below Diversion Dam or at Glenwood Bridge, downstream of Boise.

thought that would be large enough to avoid water ever going over the spillway and causing erosion. If the forecasters *over*estimated the runoff, on the other hand, they would send too much water down the river too soon, making more space behind the dam than was really needed, and the irrigators would begin the growing season with reservoirs only partly full.

Of great importance to daily operations were the "rule curves" attached to the agreement. The curves were on a complex graph and represented the engineers' accumulated hydrologic and forecasting expertise. They provided the formulas for how much water to release from each dam—given how much water was in the reservoir and how much water was assumed to be in the watershed—to keep the river flow at or below 6,500 cfs through the city of Boise. Despite the formidable technical appearance of the graph, the rule curves reflected human judgment and contained a margin of safety. The question was: Who benefited from the margin? If the rule curves allowed extra water out of the reservoir in the early part of the flood season, there would be less chance that the valley farmers would be flooded, but more chance that the desert farmers would have less than a full reservoir later in the spring. If the rule curves were too tight, things would go the other way—the reservoirs would be full, but the valley might be flooded with late spring releases greater than 6,500 cfs.

The authors of the *Reservoir Regulation Manual* wrote that the safety factor was "in favor of flood control."[7] Although there has been lively debate among Corps and Bureau engineers about the true nature of the safety margin, the general consensus now is that it did not favor flood control. It favored irrigators, not valley farmers who suffered the flood damage.[8]

The memorandum of agreement went on to deal with the order in which the Boise's three dams were emptied and filled. Since Lucky Peak was the last control point before water was diverted to irrigation, water could be passed from the upper two dams with some flexibility without being lost from storage. The agreement specified that after the flood season, water would be transferred from Arrowrock to Lucky Peak in order to keep Lucky Peak full as long as possible. The reason was that since Lucky Peak was closest to the population center of Boise, a full and level pool would provide the most convenient recreation opportunities. Once Lucky Peak was full, irrigation releases would equal releases from Arrowrock. Arrowrock in turn would be replenished as Anderson Ranch released water for irrigation and power generation.

After the flood season the Bureau of Reclamation delivered water according to the rights and contracts of the irrigation water users. The Bureau had no other clients and no obligation to respond to requests by anyone else interested in the water. The agreement specified that any changes affecting storage rights in the reservoir system would have to be agreeable to everyone with rights in the system.

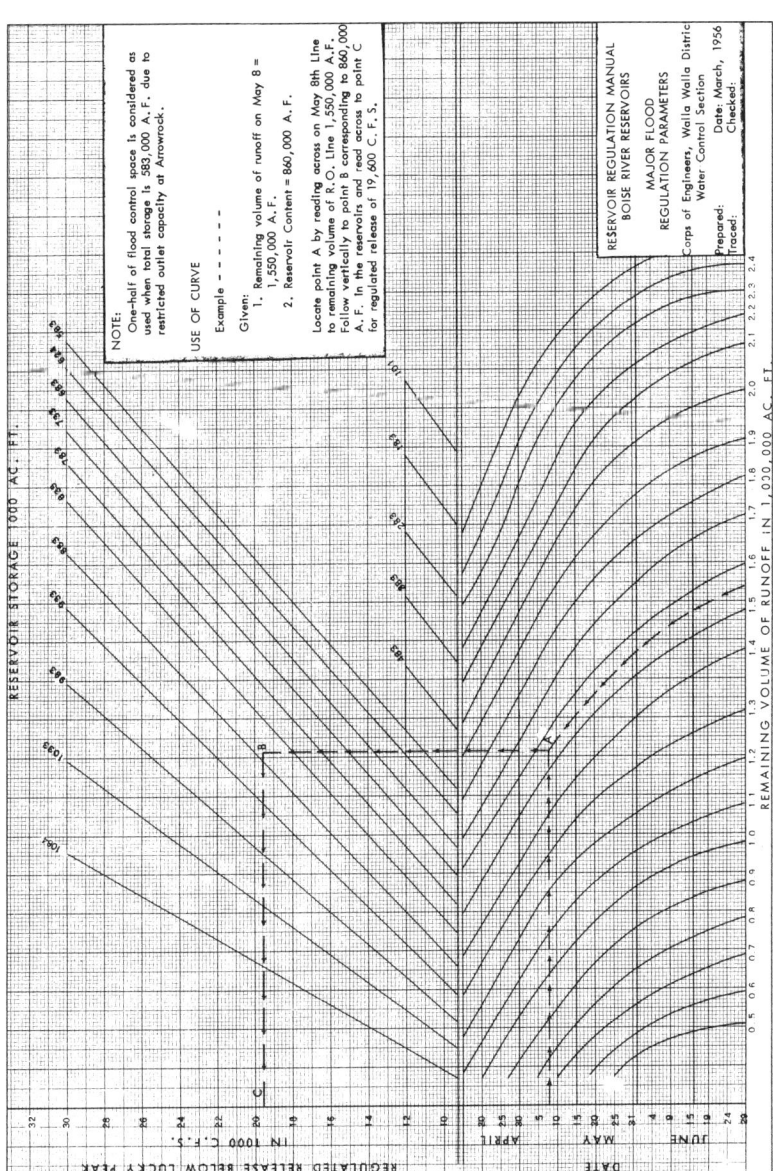

**Figure 4.2**

"Rule curves" like these are used to calculate how much water should be released from Lucky Peak, given the rate of runoff, available reservoir space, and other factors.

50

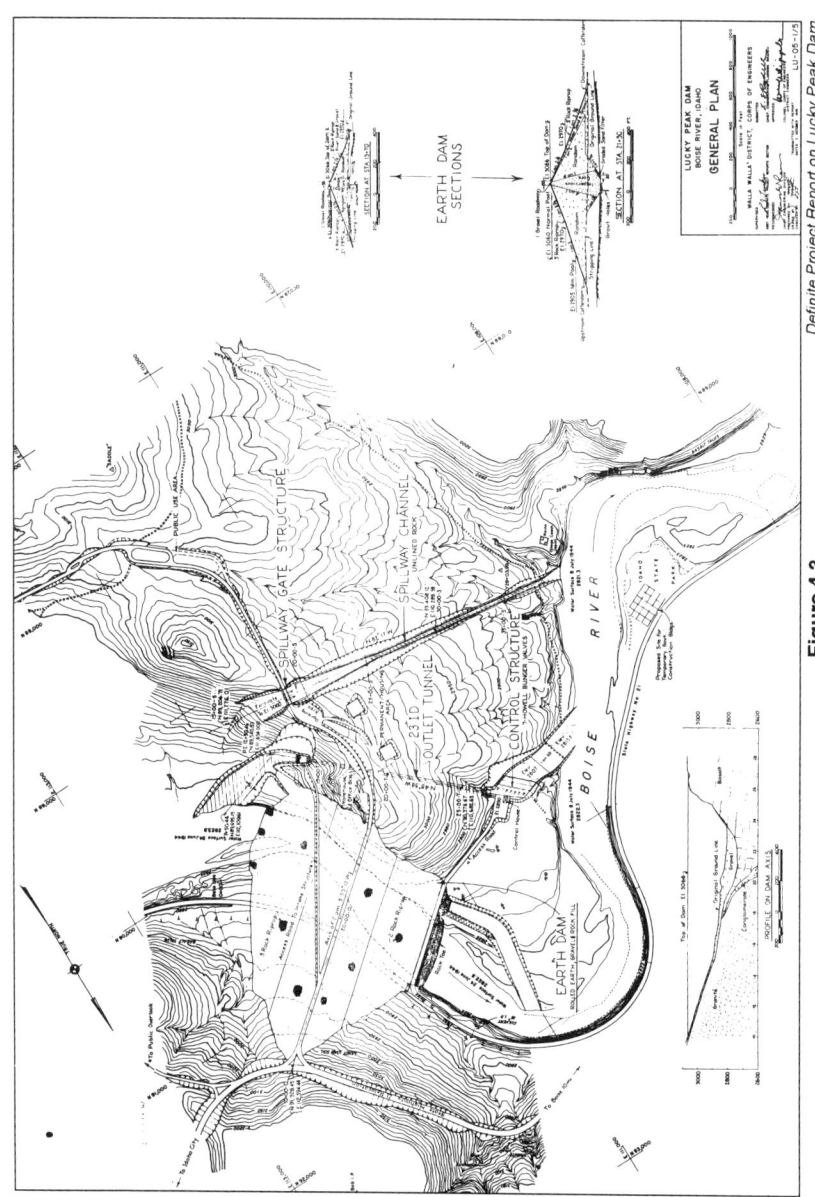

**Figure 4.3
General plan of Lucky Peak Dam**

*Definite Project Report on Lucky Peak Dam*

If a new power plant were ever installed, for example, it too would be operated "in subordination to any water rights valid under law."[9]

In summary, the one outlet tunnel, the operating deal which allowed no one to share decisions but the Bureau and the water users, and the uncertainty of the weather all set the stage for conflict when someone new wanted Boise River water. The Corps was an element, however willing or unwilling, of the conservative *status quo* and took the brunt of the hostility from the fishery interests, the flood-plain farmers, and then Boise, each in turn.

## Fish and Game Fight for Minimum Flow

In 1949 the Idaho Fish and Game Department and the U.S. Fish and Wildlife Service went on record with their desire for an assured minimum flow for a fishery in the Boise River. The Corps told them the minimum winter flow would be 80 cfs.[10] Fish and Wildlife therefore made an analysis of fishery impacts on the basis of an 80 cfs minimum flow, unaware that the Corps could not possibly supply this minimum flow each and every day of the year because of the dam's sole outlet tunnel and the fact that the tunnel had to be closed off periodically for repair and maintenance. Most likely, the Corps itself had not thought through the consequences of maintenance on minimum flow. In general, the Corps thought fishery productivity would benefit from the more even flow of the river that Lucky Peak would bring about.[11] The U.S. Fish and Wildlife analysis, assuming the 80 cfs flow, agreed that Lucky Peak would provide net benefits to the fishery.[12]

Also in 1949 the U.S. Public Health Service recommended that the river should operate with a "high surcharge followed by sharp drawdown" in order to help cleanse the river. The Service hoped that such a routine would strand driftwood and "floatage" so it could be piled up and burned to help control encephalitis and malaria. But the Corps did not incorporate this idea into the operational routine.[13] Such a routine might also have helped to move silt and other accumulated wastes down the river.

After construction began in November 1949, the Fish and Wildlife Service became concerned that the 80 cfs used for analysis was not a solid figure. Throughout the construction period, the Service importuned the Corps for clarification. In 1951 Fish and Wildlife's Acting Director Paul Quick asked bluntly what measures the Corps would carry out on behalf of the fishery.[14] Colonel W.H. Mills, the district engineer in the newly formed Walla Walla District, (which had been carved in 1949 out of the Portland District partly because of a desire that management operations be closer to Lucky Peak), said that the 80 cfs appeared to be in excess of customary stream flows below Arrowrock during winter months. He wrote that it was not up to the Corps to handle the problem, but the Bureau and water users:

The question of augmenting these flows in order to supply the minimum of 80 c.f.s., is a problem which must consider water allocations and be studied in cooperation with the local water users and the Bureau of Reclamation. The anticipated reservoir regulation will provide for this flow during the spring, summer and early fall months.[15]

But not the winter months. The first river shutoff made the technical problem very clear. When the watermaster informed the Fish and Game Department in 1954 that the Corps would shut off the river for 75 to 90 days in order to complete construction work on the outlet works penstock (the large pipe for conducting water from the reservoir through the dam), Idaho Fish and Game Director Ross Leonard reacted immediately. With Idaho Assistant Attorney General J.R. Smead, he wrote to Colonel A.H. Miller, the Walla Walla District Engineer who replaced Colonel Mills, detailing the tragic results of a dry river:

> This means, inescapably, the killing of all the fish in that part of the Boise River, not by the hundreds but in terms of thousands. Also, it means the destruction of duck and goose hunting on and along the river, both of the local ducks and of the northern ducks and geese that come in during the hunting season. That hunting area is frequented by hundreds of hunters . . . Further, there are along the river many water users whose rights . . . have absolutely first priority for the use of water for domestic purposes including stock water.

The letter concluded with a list of threatening legal citations and reminded the Corps that such shutoffs had never been necessary during either Arrowrock or Anderson Ranch construction. Leonard insisted on a face-to-face meeting to go over the situation and discuss minimum flow.[16]

The Corps, the Bureau, the water users, sports organizations, and others met on September 30, 1954, each expressing with a greater or lesser degree of cordiality ideas about solutions for the problem.[17] The group discussed the chlorine content of the sewage discharge at Boise's Lander Street sewage treatment plant and how it might be reduced to zero while the river was shut off. The state Health, and Fish and Game departments agreed to monitor the situation to prevent chlorine from killing fish where there was not enough dilution.[18] Splash boards would be placed on Diversion Dam so that the small reservoir behind it could be filled to capacity before the shutoff, providing at least some flow during the long closure. The Corps promised to release water in an emergency if at any time a loss of fish was likely.[19] None of these measures solved the basic problem, of course, which persisted until it was solved with a second outlet tunnel 30 years later.

Nevertheless, the Corps representatives sympathized with the fishery problem, recognizing that it was partly a political problem: water users did not wish to give

U.S. Army Corps of Engineers
**Figure 4.4**
**Colonel A.H. Miller, Walla Walla district engineer.**

water to the fishery at the expense of irrigation.[20] In response to a Fish and Game appeal for help, the engineers tried to supply a technical solution. They went to their calculation tables and determined that an all-year minimum flow of 80 cfs would not interfere with dam operations for either flood control or irrigation. Colonel Miller wrote to the state reclamation engineer with these findings and said that, provided there was water in the reservoir, the Corps was willing to modify the operating schedule for Lucky Peak Dam to accommodate the fishery.[21]

However, State Engineer Mark Kulp and the irrigators did not wish to compromise any water that could be used for irrigation. Kulp replied that the river had often run dry before, that he did not wish to make any agreements for the release of a specified flow, and that the "status quo with voluntary cooperation" among all parties would attain "satisfactory results."[22] The 30 cfs leak that the Corps had so carefully prevented in designing the dam probably would have looked very good to the Fish and Game Department.

With each year of operation bringing extended river shutoffs, Fish and Game kept pressing for a minimum flow.[23] In 1957 the department filed an unprecedented request for a water right for 100 cfs in the Boise River. The application provoked vigorous protests from the irrigators, who rushed to assemble arguments against it. There were no provisions in the laws, they said, for the allocation of water to permit "fish to swim as they please." Neither the Idaho state constitution nor any other law recognized fish protection as a beneficial use. Besides, the entire winter flow was needed to fill Lake Lowell and there was not enough left for fish. Letting water just flow down the channel and out of the state, preventing anyone else from appropriating it, was clearly a "waste" of the water.[24]

To these protests Fish and Game answered that fish had "prior rights" to the nonconsumptive use of water for propagation. These rights had existed long before statehood and could not be superseded by irrigation. Failure to recognize the problem of encroachment on these rights could lead to the "ultimate and total destruction" of the fish.[25] A representative of the Corps attended the hearing but remained silent.[26] The Corps' previous offer to accommodate the fishery had derived from its technical analysis, but the Corps did not bring it up again at the hearing, undoubtedly reluctant to enter into a local political discussion.

Several weeks later State Reclamation Engineer Kulp denied the Fish and Game application. He avoided the issue of whether fish had "rights" and relied on the technical principle that Fish and Game had failed to show that water actually would be diverted from the stream for its beneficial use, diversion of water being one of the rules of water appropriation in Idaho. In the past, water had been allowed

Boise Public Works Department
**Figure 4.5**
**Idaho Fish & Game biologists electrofish the Boise River to count populations of game versus "trash" fish.**

to benefit fish only where hatcheries diverted it for nonconsumptive use and then returned it to the stream, where it was available for further appropriation.[27] The issue of fish "rights" would have to be left up to the state legislature.[28]

After this rejection Fish and Game then set about trying once again to make a cooperative arrangement for the use of unallocated storage water in the dam during the winter months. Since the Corps controlled water releases after January 1, Fish and Game did not see why the Corps could not begin the releases a little earlier in order to benefit the fishery. There was no need, Fish and Game officials felt, to wait. The Bureau, the irrigators, the Corps, and Fish and Game spent the next few years discussing just how many acre-feet of water from the reservoir would actually be needed to provide for a minimum flow adequate to prevent the stranding of fish and to dilute the chlorine-treated effluent from Boise's treatment plant.

In 1958 Congress passed a critical amendment to the Fish and Wildlife Coordination Act of 1946, which authorized the Corps to consider the impacts of its operations on fish and wildlife, allowing project modifications to prevent damage to fish resources.[29] As it pertained to Lucky Peak, the act enabled the Corps to consider changes benefiting the fishery without changing the memorandum of agreement.

At last in 1963 the Bureau issued a permit to Fish and Game for the use of 50,000 acre-feet of flood-control storage space (when it was available) in Lucky Peak for low-flow augmentation. The Corps considered this *de facto* modification of the memorandum of agreement as "mitigation" for pre-Lucky Peak fishery conditions.[30] The irrigators tolerated the Fish and Game permit because the water came from the flood-control operating space. Also, they knew that in low water years this water would be assigned to them on a priority basis anyway. There was a big difference between a cooperative agreement for minimum flow and a recognized legal right to the water. The Bureau was still thinking about the Mountain Home Desert, and others had not given up the dream of a reclamation project there in the future. At a meeting in June of 1957, the Boise River watermaster had argued against providing winter flow for the fishery because "the Mountain Home Desert project will take all of the water."[31] If those new lands were to be opened for irrigation, the water would have to be kept withdrawn from appropriation so that a future irrigator could have the water rights.

The Corps fine-tuned the river so that flood storage water up to 50,000 acre-feet, formerly held until after January 1, was released earlier, giving Fish and Game some hope now for a small fishery in the Boise River. It was a small adjustment in water use at a time when irrigators regarded the water in rivers exclusively for economic development and when state laws did not endorse any contrary view.

Fish and Game still had to cope with river shutoffs, but these became shorter, less frequent, and less threatening. Although the irrigators surrendered no assets or rights because of the Fish and Game agreement with the Bureau, they profoundly regretted the acknowledgment of other water uses. Many years later the Corps and Bureau would formally recognize the acknowledgment in a new reservoir regulation manual.

## Valley Farmers Fight Flood Damage

Once the Corps began to operate the dam, valley farmers noticed—with a great deal of surprise and anger—that the Boise River continued to damage their diversion works, levees, and land. Instead of the rapid crest and fall of former floods, the river ran at bank-full capacity (6,500 cfs) steadily for weeks at a time, exerting a powerful erosive pressure. Farmers took to writing letters. One wrote to Senator Henry Dworshak: "Since building Lucky Peak Dam the river has changed more in our area than in the past 20 years. Surely there must be some means of getting an appropriation to help out financially in this pertinent situation."[32] A group of farmers complained to the Corps Walla Walla district engineer:

> Since the construction of Lucky Peak Dam, we feel that with careful forecasting and planning, there may never need to be high water again along the lower river; however, the dumping of this excess water before or after the peak flow can and does cause just as much bank erosion as natural flooding, so we must go right on maintaining these banks and channels for our own protection.[33]

This problem was no surprise to Corps engineers, who had known all along that the dam would not be a final solution. They had written in a public information brochure in 1955 that the sustained releases of 6,500 cfs could create erosion and channel changes. Bank stabilization and levees would still be needed.[34] The District Office of the Corps candidly agreed with the complaints and replied that controlling the runoff into the Boise River had completely changed its regimen and caused a new set of problems.[35]

The Corps dealt with the worst of the "pertinent situations" by authorizing emergency repairs under the 1950 Flood Control Act.[36] It justified such repairs as "advance flood fights" to prevent "imminent channel changes" from threatening farms and to prevent having to build new levees up and down the river.[37] In addition, the Corps completed a detailed study of the new flooding problem—also authorized by the 1950 Act—and published it in 1958 as *Downstream Channel Requirements*.

In looking at the "sustained high flows" the Corps analysts decided that the solution was to create a larger channel in the river. The Corps admitted that the way the river was being regulated did not provide satisfactory protection for valley landowners. The river managers knew that there would be a great deal of political uproar if the Corps ever were to flood the valley when storage space remained in the reservoirs—even if it would prevent later flooding with worse consequences.[38] The Corps concluded that a channel of 10,000 cfs capacity would be the best solution, taking into account costs, benefits, and other factors.[39]

Essentially the Corps felt that acceptable flood control really was not feasible with a flow release of 6,500 cfs. Dam operators would occasionally need to release higher flows, and when they did, water would damage valley farms, just as in the past. The difference was that the Corps of Engineers was at the controls, not the ungovernable force of Nature. Although another alternative might have been to amend the memorandum of agreement and the rule curves, this was not considered. To do so would have involved the irrigators in a renegotiation.

Giving the Boise River a capacity of 10,000 cfs would require building a comprehensive system of levees and revetments, straightening the more extreme bends of the river, clearing the channel, removing bars—exactly the measures that had not been considered cost effective in 1944 and 1949.[40] The Ada and Canyon County commissioners would have to sponsor the project and guarantee right-of-way for the improvements. The Fish and Game agencies responded by suggesting that if precautions were observed, such as taking fill material from places other than spawning beds in the stream, deleterious impacts might be minimized.[41]

By the time the Corps contracted an engineering firm in Boise to prepare the design specifications for the work, the Canyon County commissioners withdrew their sponsorship, finding they had other priorities for local funds.[42] When the design specifications became public after June 1963, local hunting and fishing organizations reacted with an outraged "No!" to the proposal. Under the leadership of Stanley M. Burns, president of the Ada County Fish and Game League, they vented their years of frustration with the Corps for not being more helpful with the minimum flow issue. Burns blasted the Corps, using such epithets as "stealthily," "impending disaster," "entirely needless," "pork barrel," and "busy work." He envisioned a river encased in a concrete ditch, a sluiceway, dead. Newspaper headlines called it a "dredge plan."[43] About Lucky Peak, Burns said:

> People still wonder why it came to be built—but the declared purpose of its fine construction was to scientifically regulate the river flow so as to finally prevent all downstream flooding. So why then another flood control [project] below our flood controlling dam? The plain fact is that Lucky Peak is not being used for its designed purpose of flood control. Its flow of water is being usurped and is being used at whimsy to further regulate only irrigation water (all for free) to the various districts."[44]

Reaction followed. Representatives of the Boise Project Board of Control tried to defend the Corps against accusations that its engineers had been "regulating themselves up a new construction job."[45] Boise mayoral candidate C. Leo Holt took the other side and tried to make the matter a campaign issue. He challenged the Ada County commissioners to withdraw their sponsorship.[46] Eventually they did, reluctant to take on both hunting and fishing interests and the Fish and Game Department. The commissioners also found it difficult to obtain rights of way for the levees. The Corps ended up placing the project on inactive status in 1967.[47]

But the flood damage with 6,500 cfs releases continued. The Ada County commissioners suggested that valley landowners form single-purpose flood-control districts, which they did. Under existing Idaho law, this enabled them to raise their own local revenues and qualified them to sponsor local Corps projects. The leaders of new Flood Control Districts #10 and #11 promptly asked the Corps to reactivate the project.[48]

The Corps did a second study, the *Levee Restudy* of 1976. This one more sensitively included alternatives for "set-back" levees in response to fishery objections.[49] These would allow the river a wider berth for shallow flooding. But the farmers saw the issue differently and balked at giving up a quarter to a half mile of land for the sake of "duck habitat" at their expense. The benefit/cost ratio did not work out for any of the alternatives unless Canyon County lands were included in the work. The landowners and duck hunters could not come to an agreement, so the Corps gave up the project once again. Prevention of flood damage still was not a goal for which either the local political leaders or the Corps wanted to fight. The 6,500 cfs—and higher—releases continued.

Valley farmers finally managed to get the attention of Idaho Governor Cecil D. Andrus. In the early 1970s the Boise watershed was in a wet cycle, but dam operations continued as usual. Outraged farmers watched bulging spring releases damage their property after winters of miserly flows. In 1974 the governor ordered the Idaho Department of Water Resources to make a comprehensive review of the Boise River system, particularly the low-in-winter and high-in-spring flow-release pattern.[50] The report found much room for improvement. Its key finding was that the safety margin in the rule curves favored irrigation to the detriment of flood control. Naturally it recommended that the rule curves be changed in order to reduce the flood risk. It took 11 years before the *Reservoir Regulation Manual* was officially revised, but new rule curves went into use in 1976. From then on, the rules provided a slightly greater chance that reservoirs might not completely fill, but they reduced the risk of flooding.[51]

## Boise Seeks Second Outlet Tunnel

For most of its history, Boise's population grew at a steady, slow rate. Then during the decade of the 1970s both the population and the economy boomed. Stimulated by expanding corporate offices, a growing university, and new high-tech industry in town, the annual rate of growth during the decade averaged over 4 percent.[52] Corporate expansion and the demand for over 2,000 new houses each year fed a healthy construction industry and created pressure on all public services, including sewage treatment. Boise urgently needed Environmental Protection Agency (EPA) funds to build a new treatment plant and applied for a grant.[53]

One of the main missions of the EPA was improving the cleanliness of the nation's rivers and streams. For EPA officials, clean water meant clean every single day of the year, every year, no matter what. They quickly understood that the single outlet tunnel at Lucky Peak meant that there would be days, perhaps weeks, when the city could not rely on a minimum flow due to shutdowns at the dam. For Boise, this translated into a serious technical choice: achieve clean water standards by somehow guaranteeing a diluting flow and build a secondary sewage treatment plant or, if a diluting flow could not be guaranteed, install tertiary treatment—an extremely costly alternative.

EPA wasted no time being subtle. Without minimum flow, it said, secondary level sewer projects "may not be approvable for Public Law 84-660 grant assistance."[54] William Ancell, Boise's director of public works, was reluctant to spend the extra millions of dollars for a level of treatment "so ridiculously high . . . that we could discharge into a dry river bed and still have water clean enough for somebody to fish in," when "relatively minor refinements in the operation of Lucky Peak might accomplish a minimum flow."[55] Ancell, new to Boise in 1972, began making inquiries about Lucky Peak operations, unaware at first that he was "scratching an old wound" by bringing up minimum flow.

Ancell contacted the Bureau, the Corps, and the Idaho Water Resources Department (successor to the state reclamation engineer), stirred up the discussion, and kept enough pressure going so that reports, studies, and alternatives began to be developed. The idea of installing a second outlet tunnel at Lucky Peak took awhile to gain a footing, but eventually became an accepted solution to the problem.[56]

The progress of the idea—and the success of the Public Works Department—is made evident by comparing the "draft" and "final" environmental impact statements (EISs) that the Corps prepared for Lucky Peak. Although the dam had been completed before the passage of the National Environmental Policy Act of 1969, which required federal agencies to prepare EISs, the Corps decided that each of its districts had to evaluate the impacts of ongoing operations and maintenance procedures at its projects.[57] Walla Walla completed its draft EIS on Lucky Peak in 1973, before Ancell's agitation had made much impact anywhere. It concluded that the dam should continue to operate as it had been—with no changes. Water quality problems "could not be avoided" and a higher level of treatment could be installed by the city of Boise, said the Corps.[58] Incensed, Ancell wrote to the Corps district engineer, "The dam never should have been built without some alternate means for water to pass the dam so that a continuous flow would be assured at all times." He suggested that such an alternate means be completed immediately. It would be far cheaper than Boise having to provide tertiary treatment. Ancell demanded a meeting with the Corps to discuss the issue.[59]

The EPA continued to press the city: "We think that a commitment from the Corps of Engineers to take a positive approach to resolving the problem must be prerequisite to the final awarding of a . . . grant."[60] Ancell's requested meeting took place, and the upshot was that the Walla Walla District Office quickly got permission from its North Pacific Division Office to evaluate more seriously the outlet problem.[61] The Corps incidentally decided to combine the study with a consideration of the dam's hydroelectric power potential, partly in response to the nation's energy crisis earlier in the 1970s.

Corps officials, Ancell, and others discussed a variety of alternatives—pumping water over the dam during a shutoff or creating a flow by pumping groundwater into the river—but the Corps finally announced in 1975 that it supported building a second outlet tunnel and installing a hydroelectric generating plant using the existing outlet tunnel.[62] The final EIS incorporated these proposals as recommended changes in Lucky Peak operations. Boise—and EPA—were at last satisfied.

Other participants in the discussions, namely the irrigators and the state director of Water Resources, never did feel that "dilution is the solution to pollution."[63] Ironically, these economic development interests had the more exalted environmental quality position, although it was not motivated by an interest in clean water. In the arid West, using stream water to dilute sewage was poor management, they felt, when cities could feasibly substitute tertiary treatment plants for dilution water. Irrigators, on the other hand, had no substitute for water.

Nevertheless, Boise Mayor Dick Eardley, Governor Cecil Andrus, and Senator Frank Church all pressed for congressional authorization for a second outlet tunnel at Lucky Peak. They succeeded. One of the last sections added to the Water Resources Development Act of 1976 included authorization for a seven-foot diameter tunnel. President Gerald Ford signed it into law on Oct. 22, 1976.[64] The next step was to develop and finance the project. The usual expectation when hydroelectric power became a component of a federal dam in the Pacific Northwest was that the Bonneville Power Administration (BPA), a public power producer and distributor, would control the power.

Despite their reservations about using water for dilution of urban sewage effluent, local irrigators were not entirely unhappy with plans for a second outlet tunnel. They had worried for years about the possible consequences if an emergency shutoff were to interrupt the supply of water during the peak of the hot summer growing season. However, they watched warily as the Corps made preparations for a new hydroelectric plant, concerned about new operating changes that might affect irrigation prerogatives. They knew that if the Corps built the plant, BPA would market the power, and there would be no advantage in this to the irrigators. They began to think about other alternatives and came to see in the proposed power plant an altogether new opportunity. They took action.[65]

Just before an October 1976 public hearing on the flow maintenance issue and before the Corps released the final EIS, the Boise Project Board of Control announced its proposal that the irrigators, not the Corps, should construct the power plant. The board listed several reasons why this was a better idea. First, the Lucky Peak reservoir had backed up to and flooded the lower face of Arrowrock Dam, which had been planned for the future installation of a power plant but had never been realized. The irrigators believed they had been given the rights to any profits

# Conflict Over Boise River Water

*Lucky Peak Power Plant Project*

**Figure 4.6**
A rare sight: discharge from Lucky Peak Reservoir is switched from the new auxiliary tunnel to the Corps of Engineers' tunnel on July 1, 1988. The old tunnel is used for hydroelectric power production.

from power sales at Arrowrock, and Lucky Peak's reservoir had caused a loss of available head. The irrigators would use the income to modernize and improve the irrigation system and help reduce the impact of rising costs of operations and maintenance. Second, federal construction meant delays; private industry could build this hydroelectric plant faster. Federal authorities would market the power outside the region, but the irrigators would see to it that the power stayed home. The irrigators deserved a "small measure of return" for all their investments in Arrowrock and other irrigation works. And finally, they did not want any power plant on the Boise River to operate to the detriment of irrigation. Having irrigators control the power plant would guarantee that this would not happen.[66]

After initial resistance, the Corps acceded to the irrigators' plan. Idaho Senator James McClure supported the irrigators and charged the Corps with "holding hydroelectric power development at Lucky Peak for ransom to the detriment of power consumers in Idaho."[67] Congress quickly passed an amendment to the second outlet authorization, supported by both Idaho senators, allowing an enlargement of the diameter from seven to 23 feet and permitting the Boise Project Board of Control to install the power plant.

The solution had something in it for everyone. The technical/engineering barrier to a reliable minimum flow would now be eliminated. The Corps of Engineers acknowledged that the missing second outlet tunnel had been a "design

**Figure 4.7**
William Ancell, right, with Idaho's U.S. Senator Jim McClure, center, and Boise Mayor Dick Eardley at the 1976 dedication of the West Boise Sewage Treatment Plant.

deficiency" in Lucky Peak, a matter at last recognized in the final EIS in 1976. The irrigators would construct the power plant and control the sale of its power.[68] And EPA was happy to fund Boise's new West Boise Sewage Treatment Plant.

Boise, like the Fish and Game Department and the flood-plain farmers, had entered the sanctuary where water-use decisions were made. Although the city, like the fish, had not acquired any water rights, the second outlet tunnel was a concession to the demands of urban dwellers for reasonable sewage treatment costs and the use of the river's water to dilute sewage.

## Changing the Rules

With the second outlet tunnel on the way, it was time to revise the old memorandum of agreement and the *Reservoir Regulation Manual*. This time, the negotiating table was much more crowded than it had been so many years ago. Boise had earned a place there with Idaho Fish and Game and other interests next to the old regulars—the Corps, the Bureau, and the irrigators.

*Conflict Over Boise River Water* 63

Other things had changed as well. The list of water users had obviously grown longer since 1955. Each new interest had wrought a change in the way the Corps managed the Boise River. An environmental movement had swept across the country and brought new laws and new procedures for public participation and review of government decisions. This movement had placed the Corps and its works under a severe and uncomfortable national scrutiny. The agency emerged from it and became accustomed to managing its planning and projects in open forums with opportunity for wide public knowledge and participation. The revision process would be very different from the one conducted quietly in the 1950s.[69]

The Corps and Bureau hydrologists came to a mutual understanding about how to adjust the "margin of error" in the rule curves. The solution rested on an agreement that their end-of-spring forecast models had been too limited. They agreed that the models had provided a poor basis for the rule curves at the most critical part of the flood season—when late runoff threatened to overtop reservoirs that were already nearly full. As years of record-keeping had accumulated, both agencies had better data and better analysis of it. They had a computerized snow and water gauging system (the Soil Conservation Service Snotel system), each

**Figure 4.8**  *Boise Public Works Department*
**The West Boise Sewage Treatment Plant discharges its effluent to the south bank of the Boise River after secondary-level treatment.**

other's basin models, satellite imagery, and instant electronic communication— even modems at the private homes of operations managers.[70]

This new technical expertise had narrowed the range of the agencies' differences of judgment over the years. Shared insight allowed for operating compromises and pointed the way to further analysis and research in improved forecasts. The solution to the rule curve problem was to provide a larger margin for error for end-of-season forecasts and to run the risk of not filling the reservoir completely. These changes improved flood control.

The two agencies also agreed that using uncontracted water (the Mountain Home Desert water) for minimum flows would benefit flood control. Part of the unallocated space in Lucky Peak Reservoir would share a priority for refill along with other users, but part would not. This latter part, 50,000 acre-feet, would be reserved as the last space in the reservoir filled during the flood season, and would only be filled after the danger of flood had passed. There was still a proviso that water reserved for streamflow maintenance could serve other uses, should demand arise at a later time.

Peter Palmer/USDA Soil Conservation Service
**Figure 4.9**
**A typical Snotel site.**

All of these agreements were implemented with an amended permit filed with the state of Idaho for the storage of water behind Lucky Peak. The water reservations were:

      13,905 acre-feet—exclusive flood control
    111,950 acre-feet—irrigation and joint flood control
      50,000 acre-feet—Fish and Game; joint flood control
    102,300 acre-feet—stream flow maintenance, municipal
                          and industrial uses, and joint flood control.[71]

The new manual clearly broadened the range of water-use priorities. In a way, its new list symbolized the changes in the social and political landscape in the Boise Valley since Lucky Peak had been authorized. Primary uses were flood control, irrigation, stream-flow maintenance, municipal and industrial uses; secondary were power (at Anderson Ranch), recreation at Lucky Peak, incidental recreation at Anderson Ranch and Arrowrock, water quality, and the sedimentation pools. The minimum streamflow was declared to be 80 cfs, which, with the Fish and Game allocation, in normal years would be at least 150 cfs. The manual candidly noted that the goal of *optimum* flood control conflicted with all other uses of water and was not the goal of the operating plan since an optimum flood control policy would require that all the reservoirs on the river remain empty.[72] A new agreement, now termed a memorandum of understanding, directed the Bureau and the Corps to operate according to the provisions of the new, *1985 Reservoir Regulation Manual*.[73]

After 30 years of river operations, the new manual was proof that certain Boise Valley interests had wedged themselves between the Boise River and the irrigators. The Fish and Game Department had a minimum flow for fish, the flood-plain farmers had rule curves in their favor, Boise had a second outlet. But the irrigators had not lost anything. They had managed to use their political influence to acquire ownership of a hydroelectric plant at Lucky Peak, a benefit providing them with an annual income worth $2 million from power sales.[74] The Corps and Bureau engineers had done what they were trained to do—apply new forecasting and other technology to the task of running the river. They had managed to accommodate a wide group of competing constituents. Political realities and technological progress both were reflected in the new manual.

Dave Reese, the reservoir regulation manager for the Corps at Walla Walla during the revision process, observed in an interview that the manual offered a way for Corps personnel to stay out of politics. When his office receives irate phone calls or hears from some local elected official with a complaint about water levels,

**Figure 4.10**
Officials and engineers observe the commissioning test of the new auxiliary outlet structure.

he said, the staff can reply that the Corps operates under sanction of the manual and those who developed it.[75] But regardless of how Reese and the operators feel about it, the manual itself is a political document, controlling water according to the ebb and flow of adjustment and compromise in the Boise Valley. The irrigators are no longer alone, although none of the changes affected their legal and political birthrights. The next 30 years may witness somewhat more revolutionary changes in legal arrangements if water demands arise for which technological fixes are insufficient to make a political reconciliation possible.

**Figure 4.11**  *Lucky Peak Power Plant Project*
The Lucky Peak Power Plant Project sits below the dam and just behind the now-quiet flip buckets.

# Chapter Five
# The Greenbelt

The early story of flood control on the Boise River focused mainly on the water in the river—how much or how little flowed in the channel, what time of the year it was released, and how the system of dams and reservoirs accommodated irrigation, flood control, minimum flow, and recreation. As the decade of the 1960s began to unfold, the scene shifted to the growing urban area of Boise and the land next to the river. Although the U.S. Army Corps of Engineers had predicted in 1949 that the complete harnessing of the Boise River would permit a more profitable use of flood-plain land in the city than what was possible before Lucky Peak, little happened for nearly 25 years to indicate that the prediction would come to pass: gravel operators continued hauling gravel from pits next to the river; cattle grazed in riverfront pastures; lumber yards and food processing plants contributed to the vaguely industrial and definitely polluted character of the river. Few property owners considered placing office parks or homes in the flood plain.

Nevertheless, the flood plain eventually did become more profitable for property owners and developers, but it was after a sequence of events that the Corps forecaster probably would not have imagined. The dynamic of urban change on the banks of the Boise began with the public's recreational interest in the river, not the profit motive of property owners. Public interest in turn brought about the need for a public land use policy. As new attitudes replaced older ones, policies changed. In the same way that the Corps' operations manual for the reservoirs reflected the emergence of new interests, the city's changing policy statements denoted shifting values in the community. As it turned out, Lucky Peak was not the final word in harnessing the river; there were many more flood-control structures to come.

During the 20 years after 1955, when Lucky Peak began operating, the process of change took two major forms. During the first 10 years, the aesthetics of the river changed, helping stimulate new community attitudes toward the river. In the second decade, Boise launched a greenbelt park system along the banks of the river.

## Boise River Aesthetics, 1955-1965

Many Boise residents noticed pleasant changes in the Boise River after Lucky Peak began operating. Charles Hummel, who grew up in the 1930s and '40s in a house not far from the river, recalled that before Lucky Peak the main stream was not particularly appealing. It had "a gravelly scarred-out appearance." But afterward the river began to improve:

> The river has developed significant heavily wooded sandbars and islands that didn't exist prior to Lucky Peak. And for that reason, in some ways the river has become more scenic. Also there is, I think, a heavier vegetation along the main bank than there was. Prior to Lucky Peak there was no interest in floating or rafting the river because it was either too low to be interesting or too high and wild to be safe."[1]

Because the annual flood no longer scoured the channel, vegetation had a chance to take hold on small islands and gravel bars. Tree branches and shrubs dipped gracefully over the water, making it seem like the river's banks had narrowed and closed in somewhat on the river.

Another aesthetic impact was not so positive. The sluggish winter flows and river shut-downs were an annual affair, although none lasted as long as the first four-month closure in 1954.[2] Low flows meant less water to dilute the chlorine-

*Susan M. Stacy*

**Figure 5.1**
**A corridor of trees and shrubs grew up along the banks of the river. Here, summer rafters drift towards Ann Morrison Park.**

*Susan M. Stacy*

**Figure 5.2**
This small island forming in the channel of the Boise River anchors vegetation and, without a scouring flood, will probably continue to grow larger.

*Boise Public Works Department*

**Figure 5.3**
Lander Street Sewage Treatment Plant went on line in 1950 and discharged chlorine-treated effluent into the Boise River.

treated discharge of the sewage treatment plant, a hazard to trout. Low flows also encouraged the growth of slime and sludge at the many places along the river where food processors and dairy operators dumped their waste products of blood, milk, and offal.

From the point of view of water quality, the river's new flow regime could not have been worse. Waste discharges peaked when river flows were at their lowest—in the fall and early winter months. To further exacerbate the problem, the missing scour action of the spring flood no longer flushed the previous year's accumulation of waste. The 6,500 cfs operating objective could not do what former flows of 10,000 cfs to 20,000 cfs had accomplished in cleansing the stream at least every other year. In short, Lucky Peak created a river in which the banks became more attractive, but the water flowing by them became much less so.[3]

Idaho Department of Fish & Game

**Figure 5.4**
**Biologist William Webb of Idaho Fish and Game examines fish killed by Boise River pollution in 1971.**

In 1959 fishery and health authorities considered the Boise to be one of the two worst cases of stream pollution in the state.[4] Pollution problems, so closely related to low-flow regimes, were not greatly improved by 1965. The city of Boise had stopped pouring its raw sewage into the river in 1950, and smaller towns downriver had also begun to treat their sewage. But treated sewage brought other problems. In a survey of water quality conditions along the river in 1964, William Webb, writing for the Fish and Game Department, reported that the increase in detergents going through the treatment plants "has resulted in mountains of foam entering receiving waters from these plants." Food processing plants, generally located out of range of sewage collection systems, still dumped their loads of "grease, potato peelings, beet pulp, paunch manure [offal], blood, dissolved sugars" and other unsavory material into the river. Some fish propagated in a few of the Boise's side channels, but the fish population generally consisted of "trash" fish like suckers, carp, and squawfish. Bass appeared occasionally, and whitefish "decreased drastically" because it spawns in the late fall, when the river flows were lowest.[5]

When approaching industrial polluters, health officials had little to back them up in their efforts to control pollution. They mainly relied on appeals for better public relations. State law provided no defined water quality standards or a required degree of treatment for discharges to streams. Food processors were reluctant to spend money on settling ponds or treatment methods that they were not sure would work. Public health employees also felt that they were not likely to get legislative support if pollution control meant that agricultural interests would be coerced.[6]

The worst slime beds lay downstream to the west of the city center, leaving more appealing aesthetics of the river upstream of town. Here the river offered new recreational temptations. In the early 1960s tubing and rafting down the river on hot summer days became a popular pastime. Meanwhile, the lake behind Lucky Peak invited boating, lake fishing, and water-skiing close to town. The Corps contracted the state of Idaho to manage picnic and boating facilities at the reservoir.

**Figure 5.5**  *Idaho Department of Fish and Game*
**A 1957 view of the fishing and boating activity at Lucky Peak Reservoir shows that it did not take long before Boise Valley residents discovered their new lake.**

Although many people enjoyed the new recreation opportunities, public and private landowners were slow to take advantage of the newly scenic Boise River. Boise Junior College, for example, which in 1940 built its administration building with a long U-shaped drive facing a riverfront road, later built other buildings around an internal mall and abandoned the U-shaped drive.[7] At Municipal Park on the north side of the river, people were drawn to the ball park and the picnic area at least as much or more than they were to the river.

The Boise Park Department invited the public to appreciate the river in 1959 with the dedication of Ann Morrison Park. Given to the city by Harry Morrison as a memorial to his wife, the park lands were overgrown with willows and cottonwoods, laced with sloughs and other soggy places. The Park Department removed most of the vegetation, retaining about a hundred cottonwood trees to provide at

**Figure 5.6**                *Boise State University*
**Boise Junior College in the 1940s. Prominent U-shaped drive behind the administration building was later abandoned. Note Broadway Bridge beyond campus site.**

least some shade until the new park plantings had time to mature.[8] Visitors could approach the river on grassy banks that sloped gently to the water's edge. The arrangement pleased the rafters and tubers, for whom Ann Morrison Park provided safe and easy debarkation from the river at a place where the floating expedition could be transformed into a satisfying summer picnic. Although Ann Morrison Park was popular, public authorities did not particularly promote further development of public riverfront land.

In the early 1960s Boise officials determined that it was time for the city to do some general comprehensive land use planning and upgrade the town's old and inadequate zoning ordinance. They hired Harold E. Atkinson, a planning consultant from California, to prepare the plan. In his 1963 report to the city council, he made several policy recommendations aimed at bringing about an orderly physical development of the city in the future. One of them pertained to the park system. He had noticed that there were several city parks with frontage on the river and many other parcels of riverfront land in public ownership. He suggested that the city "acquire land along the Boise River so as to create a continuous green belt of public lands stretching along the river throughout the entire length of the community."[9] He pointed out that because Boise was the capital city of the state with an economy based on trade and commerce, it "has more than the usual need and opportunity for parks and green areas." He suggested that "physical enhancement is a particularly worthwhile community goal."[10]

**Figure 5.7**  *Idaho Power Company*
Irrigation lines are installed at the future Ann Morrison Park, under construction in 1958. Most of the natural vegetation was cleared and replaced with turf and new trees.

**Figure 5.8**  *Susan M. Stacy*
Several years later, streamside vegetation reaches into the river adjacent to Ann Morrison Park, but a cleared access to the park remains.

The idea struck a chord with local officials and others, some of whom had been thinking about a recreational trail system along the river. The Atkinson report stimulated concerted action. After Atkinson left town, the planning commission, the park board, and the park department looked at land ownership and acreage data and decided to proceed.[11] They began an intensive planning process to define the linear park concept and figure out how to finance and develop the scheme. In January of 1966, the city council adopted a resolution making the "Greenbelt" an official city goal.[12]

Boiseans quickly embraced the idea for this special but costly park for public use and access to the river. "We hoped that the public could be brought to see what an asset we had in the river and that they would use the river, not turn their backs to it," said Alice Dieter, one of the park commissioners at the time.[13] Thus the Greenbelt began its life as a park project. The Atkinson Report was silent on the subject of the flood plain, as was the proposed zoning ordinance. The bright colors of the official plan map showed the Boise River as a ribbon of blue running through town flanked by the yellow of "open land and rural residential" uses in its flood plain. Development within the flood plain was not a controversial issue; policy makers and landowners alike took it for granted that open space uses of the flood plain were the only practical options for it.

## The Greenbelt Grows, 1965-1975

For the cover of the Boise Park Department's *1967 Annual Report*, an artist drew a cartoon of a little belt, complete with buckle, hatching from an egg. "Boise River Greenbelt Project is Born," said the caption.[14] The first acquisition had occurred in 1966, with the donation of a .43 acre parcel by the Taubman Corporation. Two more small donations followed in 1967.

As enthusiasm for the Greenbelt took hold, the town warmed to a debate about what the future character of this new park should be. Competing ideas flew in all directions. The park board hired a local planning consultant, Arlo Nelson, to help organize everyone's ideas into a workable plan.[15] When he was finished, Nelson's public presentation of maps and sketches stretched around three walls of the city council chamber. They tantalized the audience with images of lakes, amphitheaters, foot bridges, playgrounds—a recreational paradise that further captured the public's imagination.[16] At the same time, the city applied to the federal government for Land and Water Conservation funds, asked the Idaho legislature for surplus state land for use as a riverside municipal golf course, and made a prioritized list of the land parcels it would need to acquire along the river.

After spirited discussions the city arrived at the guidelines it would adopt for the linear park in 1968. Billboards would be prohibited, as would tree cutting and overhead utility lines. The public would have "in perpetuity unrestricted access to the river and to the special and unique forms of recreation it provides." After listing a variety of recreational uses and activities that would be compatible if adjacent to the Greenbelt, the statement added that residences and light office buildings might be "reasonably compatible" if they were separated from the Greenbelt by an adequate buffer zone. Two years later, a new item was added to the guidelines to include a function of the Greenbelt that had been overlooked in the original version, namely, that the greenbelt could use "delineated flood-plain properties on which permanent construction is necessarily restricted."[17] The city council amended the zoning ordinance in 1971 to provide for a minimum setback from the river and created the Greenbelt Committee to review any proposals that might come along.

Meanwhile, in other parts of the United States far from Boise, it had not been so obvious that construction in the flood plain was "necessarily restricted." The federal government had been spending continually increasing amounts of money for relief after flood disasters. As Congress absorbed information about the escalating cost of flood damage to the general taxpayer, it had been taking action. In 1960 it authorized the Corps of Engineers to provide information to local communities about the location and characteristics of any flood plains that existed within their jurisdictions. Corps district offices established Flood Plain Information Services Programs to do the job. In 1967 the Corps' Walla Walla office published a report for Boise City called *Flood Plain Information, Boise Idaho and Vicinity*.[18] The cover photo showed an aerial view of the 1943 flood and described it and other Boise River floods (and floods of its north side tributaries that flowed through town). A series of long pull-out sheets presented aerial photos of the flood plain and illustrated how far the river would rise in the event of a 100-year flood—and also in the event of the largest flood that might ever possibly occur. (The 100-year flood is a flood that has a 1 percent chance of occurring in any given year. The land it would inundate is designated as the 100-year flood plain.) For the first time, Boise had a concise public document showing exactly where the flood hazard was located.

The Corps suggested that the information could be used to help plan the best use of flood plains, "particularly where development has not encroached on the flood plains. With the information presented, developments may be planned at elevations high enough to avoid flood damage."[19] The Corps noted that most of the residential development in Boise was outside the flood plain, but anticipated an "ever-increasing demand" for it.[20] The report noted that the Walla Walla office of the Corps would provide, "upon request, technical assistance to Federal, state, and local agencies in the interpretation and use of the information presented."[21]

The appearance of the report did little to switch the focus of Boise's community effort away from recreation, although it raised the community's consciousness of the flood plain and explains why the "revised" Greenbelt guidelines of 1970 added a reference to it. The 1968 Greenbelt plan referred to the report also, and suggested that a flood-control zoning classification be written into the zoning ordinance "to cover areas where flooding is a hazard . . . The suggested golf course location is an example of the flood area, and a club house is a good example of construction which would have to recognize the hazard of flooding."[22] The city in 1968 was clearly preoccupied with its linear park, having as yet no thought that houses would eventually be needing flood plain zoning attention.

The park department's land acquisition program accelerated after 1970. When the city tried to purchase one 12-acre parcel for the Greenbelt, the owners refused to sell except at prices greatly exceeding the appraisals. The city council condemned the land, whereupon the property owners took the city to court. The council ended up winning the land, although paying more than it had expected, but confident that the project now had a serious future. From then through 1977, the city acquired 17 riverfront tracts totaling 72 acres. Further development included wells for sprinkling, paths, grading, landscaping, decks, a skate and bike rental concession, and a footbridge connecting the former junior college, now Boise State University, with Julia Davis Park. The state transferred 52 acres east of town to Boise for Municipal Golf Course, at that time the intended eastern terminus of the Greenbelt.[23] The land was on the grounds of Idaho's old penitentiary, but had been regarded as marginal for farming and most other uses because of its location in the flood plain.

**Figure 5.9** *Susan M. Stacy*
**Recreation facilities along the Greenbelt included viewpoints for quiet reflection and contemplation.**

*The Greenbelt* 77

Efforts to clean up the river went into full swing. Section 208 of the Federal Water Pollution Control Act amended in 1972 (Clean Water Act) gave the Corps of Engineers funds and authority to provide technical assistance to local communities as they planned alternatives to dumping untreated wastes in rivers.[24] Public agencies in Ada and Canyon counties organized a joint effort, financed largely by the Corps, to improve the water quality of the Boise River. It was one of the first times that local and federal agencies collaborated during the early planning stages to deal with such a complex problem in a comprehensive way.[25] The work led to the construction of a new wastewater treatment plant in Boise and a number of other actions aimed at reducing pollution from feedlots, meat packing plants, irrigation return flow, urban storm runoff, and other sources.

An enthusiastic citizen constituency for the "continuous belt of green lands" included service clubs, environmental groups, the newspaper, local corporations, scout troops, and thousands of citizens who rafted, tubed, partied, fished, biked, and cleaned up the river and its banks. These supporters, along with the members of the city council, began to devote themselves to the idea that the Greenbelt path system for pedestrians and bicyclists should not have to cross streets and bridges. Engineers designed path extensions under the city bridges, none of which had been built anticipating such passages. Riders and walkers would not have to cope with dangerous motorized traffic.[26]

*Susan M. Stacy*

**Figure 5.10**
**Bicyclists pass under the Fairview Avenue Bridge, one of several Greenbelt underpasses in Boise.**

The riverbed itself became a focus of civic action. In 1970 a 12-year-old boy drowned after a metal pipe, part of an irrigation diversion structure, snagged his inner tube.[27] An alarmed community, under the sponsorship of the local American Legion post and the Idaho Department of Public Lands, undertook a massive and ambitious cleanup campaign. The Chamber of Commerce coordinated the efforts of scores of corporate and government donors and a thousand individuals during the low-water season of 1971. The National Guard, the Seabees, Morrison-Knudsen, the state of Idaho, Boise State College and Boise High students, and fishing organizations did everything from removing and replacing diversion structures to picking up debris on the river bottom. The Sierra Club and Jaycees allied with the 321st (Reserve) Engineer Battalion to haul huge concrete slabs, remnants of an old flood fight, off the banks near the Americana Bridge so the park department could install sprinklers, top soil, and a path.[28] One group logged the location of remaining pollution sources. More was done the next year to remove remaining hazards, including the rebuilding of a diversion dam just below Ann Morrison Park. Scout groups and service clubs continued the practice of cleanups and enhancement projects nearly every year after that.[29]

The reluctance of some private property owners to defer to public interests on the river created conflict in 1974. The International Dunes Motel (now Shiloh Inn) built its swimming pool inside the Greenbelt setback and too close to the river, a

*Susan M. Stacy*
**Figure 5.11**
**The Greenbelt path passes between the Shiloh Inn to the left and the river to the right.**

violation of the by-then well-known zoning regulation. After the fact, the motel sought a variance to the setback, which the city's board of adjustment denied. Upon appeal to the city council, it was denied again. International Dunes took the city to district court, arguing that the greenbelt setback was unconstitutional, among other issues. The judge dismissed the complaint and the motel had to comply with the setback.[30]

By 1975 the greenbelt idea had burst out of the city limits into neighboring jurisdictions. The Ada County commissioners established the first public park in the county's jurisdiction, reversing a century-old policy against having county parks. Forced to concede that traffic had become hazardous because of random parking on the narrow road and bridge where river floaters left their vehicles, they created Barber Park on the south side of Barber Bridge where people launched their tubes and rafts into the river. The public, somewhat dismayed by having to surrender the tradition of wholesome chaos at the place, reluctantly accepted the large parking area and rest rooms as a civilized concession to the reality of 10,000 river floaters a day on summer weekends.

Other governments joined in. The state of Idaho's Department of Parks and Recreation observed the popularity of the river and the path system on its banks. The department considered how it might extend the system all the way east to its Discovery Park at the Lucky Peak Dam site itself, and later to a new state park on Eagle Island west of town near the town of Eagle. Garden City, a small enclave to Boise's west, decided to adopt a general plan goal to continue the greenbelt through its jurisdiction.[31]

Within its first 10 years, the Boise River Greenbelt was a smash hit. The city had secured several miles of riverbank for the public's perpetual "unrestricted access" to recreation, aesthetic enjoyment, and open space. The public in turn had spent thousands of hours of its labor and other donated resources on the project, an extraordinary civic investment. The river became the chief symbol of the high quality of life residents felt they enjoyed in Boise. The park board realized its early hopes that Boise would "turn to the river." Finally, housing and office developers wanted to be there too. The Boise River was the place to be.

**Figure 5.12**     *Susan M. Stacy*
**The terminus at each end of the Greenbelt shifted frequently during the 1970s and 1980s.**

# Chapter Six
# Developing the Flood Plain

Once the river cleanup and public investment in the Greenbelt had transformed the river into a sparkling urban amenity, property owners who had been content with agricultural or industrial uses of their riverfront acreages took another look. If so many people wanted to be along the river, then surely they would enjoy having offices and homes along the river. So, land developers entered the picture, and Boise spent the 1975-85 decade making, unmaking, and remaking its policies pertaining to the river. Because of the city's greenbelt setback ordinance, developers could not always get their buildings as close to the river as they would have liked, but had to compete with the public for access.

The entire decade of the 1970s was a period of substantial population and economic growth in Boise and Ada County. The old 1964 Atkinson plan withered into irrelevancy when the city surpassed Atkinson's 1985 population projection in 1976.[1] Developers scrambled to keep up with demand for new housing. The pace of growth was so rapid that the city was hard-pressed to provide it with the normal city services. Coordinating and extending sewers, public safety systems, schools, and roads was part of the problem, but developers were also pushing to take advantage of the popular Greenbelt and build in the flood plain. The growth and the problems it created cried out for a sharper planning focus and new policy directions. The city needed to equip its decision-makers with a new comprehensive plan and growth management policies in order to keep up with all of it.

## National Flood Insurance Program

In 1975 Boise Mayor Dick Eardley called a citizens group together to develop guidelines for a new general comprehensive plan. That was also the year Boise entered the "emergency" phase of the National Flood Insurance Program (NFIP). The NFIP was a response by Congress to the high public cost of flood disasters.

Passed in 1968, the program provided that the federal government would subsidize flood insurance for property owners only in communities that enrolled in the program. Enrollment required the city council to enact land-use controls designed to eliminate or substantially reduce the damage that floods could have on development in the flood plain.

If a community were interested in the program, the government gave it an "emergency" enrollment until all the local control ordinances were in place, the maps of hazard areas prepared, and all was approved by the appropriate federal agency. At first, that agency was the Department of Housing and Urban Development; later, it was the Federal Emergency Management Agency (FEMA). Meanwhile, any structure built in the flood plain had to have its first floor at least one foot above the level of the 100-year flood.[2] FEMA administered the program, and the Corps provided information about how high the 100-year flood would be in a given area and the elevation one foot above it. FEMA established minimum standards that local flood-control ordinances would have to meet in order for the community to qualify for permanent enrollment in the flood insurance program.[3]

The mayor's planning committee, which included developers and their engineers and other representatives, got around to the issue of what land-use philosophy—aside from the federal minimum standards—would guide Boise's flood-control ordinance. Boise's planning staff proposed to merge the federal minimum standards with the local goals and priorities that already had been established. The proposal sparked instant tension among members of the committee because the local goals placed a very high priority on recreation and open space, a priority that developers and their engineers noted went far beyond the federal minimum standards. The local tradition had been that it was not safe to develop in the flood plain, but these new federal standards implied that residential and business construction could be placed safely in the flood plain if it conformed to minimum standards for such construction.[4]

Naturally, developers supported the National Flood Insurance Program requirements because they permitted flood-plain land development. The developers pointed to the federal—and therefore presumably impeccable—authority behind the standards and argued that the entire flood plain should not be reserved for open space. They felt that the NFIP provided a reliable regulatory structure for safe minimum standards for flood-plain development. Elevating first floors one foot above the 100-year flood was easy to do by consulting the flood insurance maps, which in turn had been based on the work of the Corps of Engineers' 1967 analysis and later updates.[5] And finally, the federal rules also allowed for removing land from the flood plain by building levees between it and the river.

During the emergency phase, local governments and the Corps worked out a routine for reviewing local levee and project proposals. A developer would submit a development and levee proposal to the appropriate planning department, which

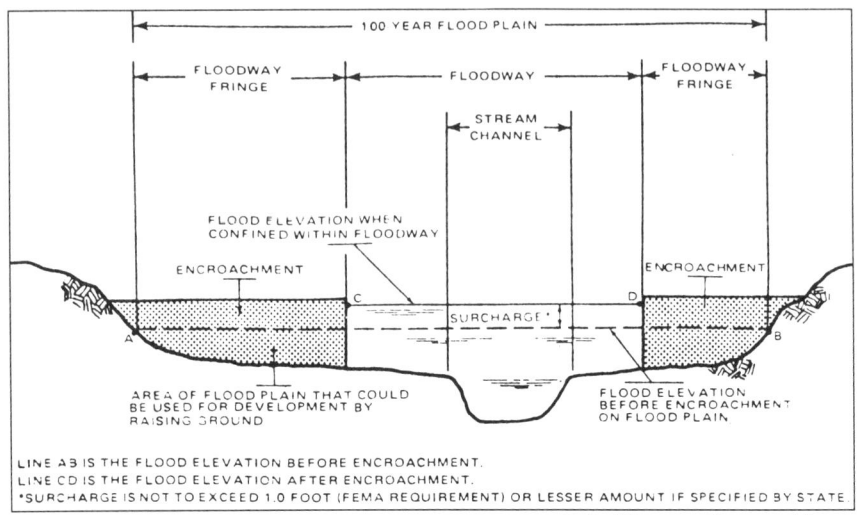

**Figure 6.1**  *Federal Emergency Management Agency*
**Schematic used by FEMA to illustrate the floodway and floodway fringe concepts.**

would in turn ask the Corps to review it. As a technical service, the Corps analyzed how the new levee might affect other property in the event of a 100-year flood, whether flood levels would be raised elsewhere because of the levee, and if so, where and how much.

The Corps performed its calculations for local communities with the aid of a computerized river model called the HEC-II step-backwater program, developed at the Corps' Hydrologic Engineering Center in Davis, California. The model and the regulations associated with it divided the flood plain into a "floodway" and the "flood fringe." The floodway included the river channel. The regulatory concept was to imagine the outer part of the flood plain filled or leveed against the 100-year flood up to the point where the flood level in the remaining channel and adjoining lands would not be raised any higher than one foot. The outer part would constitute the flood fringe and the rest the floodway. If buildings were to be placed in the flood fringe, it was critical that the floodway be left unobstructed with no buildings or fill—nothing should cause the potential flood to rise more than the allocated one foot. At the same time, nothing placed in the fringe area could be allowed to cause the flood to go any higher than one foot.[6]

Developers eventually perceived a range of options for flood-plain development, some more advantageous than others. If developers built levees to remove land from the flood-plain designation, they and their customers would not need to purchase annual flood insurance. Levees also cost less than hauling in fill to elevate the entire area on which buildings would be constructed. Developers therefore began asking for permits to build levees to remove as much land from the 100-year flood plain as possible.[7]

As Mayor Eardley's citizen planning committee stretched its work into and through 1976, the flood-plain policies provoked intense debate among developers, landowners, citizens, and planners. When the committee finally presented its recommendations, the Boise City Council adopted an ordinance in December 1976 to "receive and accept" the guidelines as the basis for amending the comprehensive plan and sent it on to the planning commission. The document contained the following pair of policies on the flood plain:

> The 100-year floodway of the Boise River should remain in a natural state as a greenbelt, wildlife habitat, or of an open space recreation nature with no structural development that will impede the flow of flood waters.
>
> Limited development should be allowed within the 100 year floodway fringes; such development should not restrict or alter the natural flow of water within the floodway nor otherwise increase the size of the existing flood plain.[8]

After the long forums and debates, the focus of Boise's river policy had undergone a subtle but significant change. The city had absorbed the Corps' concept of thinking about the flood plain as a floodway and a floodway fringe. In the process, the Greenbelt became identified as part of the flood-plain policy; the flood plain was no longer just an incidental aspect of the Greenbelt policy, as it had been up to that point. The Greenbelt idea had originated as a civic conviction that the park would enhance the city's life, not because public recreation was good flood-control policy. That a park might also make something attractive out of old gravel pits and wasted flood-prone land had been just a benefit of the park, not its major reason for being. For the first time, development in the flood plain— meaning residential and other private development—became part of flood-plain policy as well. The policy was satisfactory to FEMA. The various local factions had made their compromises: greenbelt would go in the floodway; private development in the floodway fringe.

## ParkCenter

However, the policy did not remain effective for long. Also in 1976 the Morrison-Knudsen Company, through its development subsidiary Emkay, asked Boise to annex several depleted gravel pits and other land on the south bank of the river just east of the Broadway Bridge, about 125 acres. Emkay wanted to build a high-quality office park for corporate tenants. The city owned a three-acre spur of land that jutted into the Emkay property; Emkay proposed to trade this spur for an equivalent amount of easement for the Greenbelt. The city agreed to the trade and to annex the land.[9] When combined with other land the city already owned along the river in that area, this trade provided for an extension of nearly one mile of Greenbelt. The Greenbelt seemed, with this major—and voluntary—corporate development, to have become an object of universal approval in the community. But other aspects of the development, less accessible to public view or comprehension, were about to dilute the early Greenbelt ideals.

Two months after the city council had "accepted" the principle of retaining the floodway for open space use, Emkay filed its anticipated application for the office park, to be named ParkCenter.[10] The land lay almost completely in the flood plain, the majority in the floodway itself. Aside from the discouraging policy statement the city had just endorsed, Emkay had a further problem. Because the city was a participant in the emergency phase of the National Flood Insurance Program, it could not allow Emkay to place buildings in the floodway.

The planning commission and city council proceeded with the usual development hearings. The staff report pointed out that the council's recent action on flood-plain policy only expressed an "intent" to prohibit construction in the floodway and that an actual city regulation forbidding it was not yet on the books. If the council desired to follow its "intent" to reserve the Boise River floodway for greenbelt open space, Emkay's project would have to be denied. But the staff did not recommend denial. To deal with the fact that the National Flood Insurance Program also prohibited floodway development, the staff suggested that the detailed plans be sent to the Corps for their review and comment before final city approval.[11] The council accepted the staff report recommendations and approved Emkay's development concept despite the fact that neither the city nor Emkay had any idea how the flood-plain situation would eventually be resolved. The approval, of course, reversed its erstwhile intent to reserve the floodway for greenbelt, wildlife habitat, and public open space. The permits were contingent upon a number of conditions, one being that any development in the flood plain be done only after the Corps delineated the location of the flood plain and that the project complied with the pertinent NFIP regulations.[12] Thus the council expressed its desire to have the project developed and handed the floodway problem to the engineers.

The real basis for the city's approval lay in its concerns about developing downtown Boise. For many years, the city had been implementing an urban renewal project in downtown Boise. The heart of the plan was to build an enclosed regional shopping center, a project not yet accomplished, but for which the city was reserving a significant quantity of central downtown land. A number of Boise-based corporations, including Albertsons (a supermarket chain), wished to develop new headquarters office buildings and were unable to assemble sites large enough in the downtown area or immediately adjacent to it. The Emkay site, within a mile of the city center, was large enough to offer control of the total development environment and also had the potential for mass transit service. If the city council were to foreclose that site also, it was likely that the corporations would seek land near the airport industrial area or somewhere else on the suburban fringes outside of town.[13] Given these circumstances, the council did not have a difficult choice. Either develop the Emkay site as a high-quality office park within a mile radius of downtown and the hoped-for shopping center or leave the large open area as a scarred-up exhausted gravel pit just because it was part of the floodway and lose the office buildings to a distant edge of the city. Furthermore, the city hoped to encourage more growth east of Boise, and ParkCenter would give that goal a meaningful boost. The council opted for the best hope for downtown: economic development of the floodway. The rest of the bargain included Emkay's promise to help extend the Greenbelt and to reserve, develop, and dedicate one of the old gravel pits as a public park, what is today known as ParkCenter Park.[14]

With the city's concept approval in hand, Emkay engineers confidently submitted their preliminary subdivision plat to the city, plainly showing the building lots they hoped to create in the floodway and the floodway fringe. With similar confidence they supplied their proposals for an engineered solution to the floodway problem to the Corps of Engineers. Emkay hoped to persuade the Corps that the solution could come from changing the "roughness coefficient" in the Corps' computerized flood elevation projections.* If the coefficient were liberalized, the area designated as "floodway" might not be as wide as the official flood-plain maps showed. Emkay also wanted the Corps to look with favor on its plan to raise the flood elevation levels greater than one foot in the floodway, a plan that the National Flood Insurance Program regulations could not allow. Nevertheless, Emkay argued its case with the Corps, looking for its approval. Technically, the Corps could not actually "approve" development plans, but its favorable reaction to a proposal was equivalent to the city's approval, since the city relied on the Corps for technical analysis and interpretation.[15]

Emkay's early confidence turned out to be misplaced, for the Corps refused to sanction flood-plain developments that violated FEMA regulations. Emkay was in for many months of negotiation before all parties reached an agreement about what to do. The long dialogue among Corps, Emkay, and Boise engineers commenced. The problem was particularly hard to solve because of a small waterway known as Loggers Creek, which flowed through the Emkay property. Loggers Creek was an

---

*The coefficient of roughness is a measure of friction on the surface over which water flows. Rough surfaces present more friction and slow the water down, causing it to spread more widely. Smooth surfaces allow it to pass faster and tend to reduce the land area flooded. Engineering manuals offer guidelines as to what types of surfaces should be given what coefficient, but the engineer has to use judgment in interpreting actual conditions.

old irrigation diversion that branched off from the river considerably upstream from the ParkCenter site. When the Boise River flooded, water that overflowed the river banks entered Loggers Creek, which then also overflowed and had its own floodway.[16]

The engineers all finally concluded that the only way Emkay could use all 125 acres of its site was to take it out of the flood plain altogether. To do that, they had to build a levee along the river adjacent to the site. The HEC-II model showed that such a levee would cause too much flooding on the other side of the river. To deal with this, the engineers came up with the idea of an "auxiliary channel" through the ParkCenter site to divert enough water from the river so it would not overtop the opposite bank. But there was still Loggers Creek. To prevent Loggers Creek from bringing flood water from upstream, the levee along the river would have to be extended beyond the Emkay property all the way to Loggers Creek, and Loggers Creek would have to be controlled; that is, a more sophisticated control works would have to be installed at the diversion point to prevent too much water from entering the creek during a flood. The HEC-II model showed that extending the levee this additional distance once more threatened land on the opposite bank with overtopping, so the Loggers Creek area would also have to have some version of an auxiliary channel. The original levee next to the Emkay site had sparked a kind of chain reaction generating more levees and control structures until the HEC-II model showed no more unacceptable impacts.

The Corps carefully calculated how much water Emkay's auxiliary channel would have to carry to prevent the 100-year flood from overflowing the opposite bank. The auxiliary channel was actually an easement across ParkCenter upon which no future structures or buildings would be allowed. Corps engineer Ron Barrett warned the city that if the channel were ever blocked during a flood, it could cause the river to overtop the railroad bank on the opposite side of the river. He suggested that in the event of a flood the city should watch the structure and make sure that no one prevented the flood from going through the ParkCenter development.[17]

With the plan accepted, Emkay agreed to complete the levee before December 1978 and started construction on its first building, Albertsons' new corporate headquarters, in late 1977. Development of ParkCenter continued throughout the remainder of the 1970s and 1980s. Emkay built the levee with a special concrete overflow spillway. To a casual observer, this spillway looked like a modest dip in the levee. When there was sufficient water in the Boise River, part of the flood would flow over this shallow spillway, cross the ParkCenter development, and then rejoin the Boise River downstream. Barrett predicted that a Boise River flood would have to exceed 11,000 cfs to raise the water in the main channel to the elevation of the spillway.[18]

## River Run

The Emkay floodway solution had to affect proposed developments upstream because, as Ron Barrett pointed out, the auxiliary channel would not work until upstream developments had restricted flood flows to the channel of the Boise River and had controlled Loggers Creek.[19] A major portion of the upstream property between ParkCenter and the head of Loggers Creek belonged to the owners of River Run, a group preparing to propose another high-quality project, a 118-acre residential complex, early in 1978. The project was large and ambitious, complicated by many issues other than the flood-plain requirements inherited from ParkCenter. These issues included the design of the Greenbelt and its bike path, the growing public interest in "natural" wildlife habitat along the river, and a growing market interest in riverfront homesites. Each issue was like a thread coming off its own spool at the beginning of the project, but each thread soon became tangled up with the others as the project materialized over the next seven years.

The Corps' 1967 flood-plain report showed that a substantial part of River Run's acreage was in the flood plain. Soon after the developers filed their concept application, Planning Director Gerry Unterkoefler wrote to the Corps that the city would appreciate the Corps working with River Run's engineer in order to arrive at an approvable solution to flooding problems.[20] River Run's engineer, Tony Peterson, travelled to Walla Walla and reported back that he agreed with the Corps on preventing uncontrolled flows in Loggers Creek and would be sending the Corps the necessary maps and other information.[21] River Run, concerned that the Corps would not have its calculations and conclusions ready by the time of the planning commission hearing, suggested that the commission make its approval contingent upon the Corps "ultimate judgement" about the boundary of the flood plain.[22] River Run, like Emkay before it, was confident that the Corps would quickly agree with the company's engineering data pertaining to the location of the flood plain. It expected the Corps to concur with its calculations as to where the edge of the flood plain really was, which River Run engineers felt was closer to the river than the flood-plain maps indicated.

The planning commission accepted this logic and went ahead with public hearings on the River Run project in March of 1978. Once more the city planning officials were eager to encourage development in this part of the city, particularly since River Run promised to be an exemplary project. They approved a concept plan for the project, making the approval contingent upon the Corps of Engineers delineating the flood plain and ascertaining that the project met relevant standards. The commission was unaware that by dispatching the engineers to solve the flood-plain problem, the engineers' solutions would impact the Greenbelt in unexpected ways.[23]

As part of its proposal for the Greenbelt, the River Run developers wanted to trade land with the city. Several years earlier the city had acquired land next to the river for a park with the help of federal funds. This land was on River Run's western flank. River Run wanted to use it for road access to the project and proposed that the city trade it in exchange for a strip of land along the river for Greenbelt development. Federal rules required the city to restrict use of the parcel solely for park purposes unless it could demonstrate that land of equivalent recreational value would replace it. Change in ownership or use would need federal approval. But the Greenbelt Committee and the staff greenbelt coordinator neglected this aspect of the land trade and opened negotiations with the developer without mentioning it. The trade discussions proceeded parallel with the other city departments' analysis of the flood plain and related development issues.[24]

River Run's overall concept application stated that it wanted to be the first residential development in Boise to integrate fully with the Greenbelt and its bicycle path, to provide the "largest single boost towards the completion of the bike path to Barber Park," to extend the Greenbelt along the entire river length of the project, and to provide 21 acres of public open space. The floodway would be left in its natural state, and on part of it River Run would build a small lake—Lake Heron—to separate the private development from the public park.

The application included a copy of the mutual understanding arrived at thus far between the Greenbelt Committee and the developers about the design of the

**Figure 6.2**
**Bystander observes Canada geese in Heron Lake at River Run.**

*Susan M. Stacy*

Greenbelt. They agreed that the Greenbelt would be established along the entire riverfrontage, but would not be a constant narrow strip. Rather, its boundaries would be variable in its distance to the river bank. On it would be a pedestrian pathway with benches and other stopping places situated along the way. River Run would provide a bikeway, but might locate it "partially in the Greenbelt and partially through the project away from the river." The Greenbelt Committee understood that it would "intersect the river at different points."[25] There would be "every design effort to prohibit motor access." The Greenbelt Committee was agreeable to pursuing the proposed trade of land.[26] The bike path and the Greenbelt would tie in to a point on the east end of the project where the city had another undeveloped park parcel known as Loggers Creek Park. River Run drew up its plans assuming that it would become the owner of the city's land and planned to put a major road access upon it. But all these plans were made prior to knowing the final outcome of the flood analysis. As River Run gradually learned what it would have to do to solve its flood-plain problem, its plans for the Greenbelt, the bike path, and the land trade changed.[27]

River Run engineers, convinced that the floodway at River Run was less extensive than what the Corps flood-plain maps indicated, made new surveys of the land and sent a proposal for levees and a new floodway boundary to the Corps. The Corps accepted the proposed modifications to the floodway map and also demonstrated how a few modest changes could further reduce the floodway area.[28] The Corps reminded River Run that the inlet at Loggers Creek had to be controlled to protect ParkCenter and part of River Run's property.

But there was one further problem. The calculations done earlier by the Corps for ParkCenter had indicated that the levee along River Run and a controlled Loggers Creek would shift part of the flood to the other side of the river. To prevent this, another auxiliary floodway had to be created somewhere. But where? River Run's solution was to carve the channel out of part of the land being proposed for its trade with the city. In September of 1979 River Run president Peter O'Neill wrote to the city and declared that the configuration of the Greenbelt would now have to be changed and that River Run's contribution of acreage reduced because of engineering factors affecting the construction of the flood-control works. River Run had decided to build the auxiliary channel immediately adjacent and parallel to the river. The design, with several inlets and outlets, required a number of breaks in the remaining narrow strip of land; in effect, these openings created a series of islands where there had been none before. In addition, Lake Heron would be larger and this too would reduce the acreage going to the city. Concerning the reductions in acreage, O'Neill felt that the riverfront land was more valuable than the appraisals had indicated. He did not expect River Run to profit from the land trades, he said, but "neither do we intend to unreasonably subsidize Boise City."[29]

By the end of the land trade negotiations, city officials accepted a smaller acreage from River Run in exchange for the same number of acres that the city had planned to give up all along. Instead of an intact strip of land, part of which had been of fairly substantial width, the city obtained a series of narrow islands in the river that at some places offered the public less than the ordinance-required 70 feet setback. The city realized that it would have to construct bridges across them in order to provide a continuous path and access to the river along the project. River Run agreed to pay for the bridge that would cross the Heron Lake outlet, but refused to consider financing any others.[30] The dredged channel and the River Run levee took the project land out of the flood plain and the floodway. The development unit adjacent to the dredged channels and islands was called The Island, a security-gated development for 33 single-family homesites.[31]

The original understanding about the location of a bike path also changed. The early language that the path "intersect the river at different points" evolved into a plan to place the bike path on the main road through River Run, never intersecting with the river. River Run promised to "build and dedicate to the Ada County Highway District for public use two six-foot one-way delineated bike lanes."[32]

**Figure 6.3**
**River Run apartments overlook water control works adjacent to the Boise River.**

Susan M. Stacy

Meanwhile, to its home-building clients, the developers proceeded to make representations that bicycles would not be allowed on the riverfront path system.

At last the city's neglect of the fact that the land it proffered for trade had been purchased with the Department of the Interior's Land and Water Conservation funds caught up with the negotiations. Jack Cooper took over as new park director in 1979 and realized that the federal authorities and the state administrators of the Heritage Conservation and Recreation Service (HCRS) had been left out of the loop. Worse, River Run had already constructed a road on the city's property, changing its use without federal authorization and violating the law. Cooper had the unpleasant task of advising River Run—and now its attorneys—that the HCRS had to evaluate the land exchange proposal and endorse it before the city could consummate the deal and exchange deeds with River Run. Exasperated with the delay and in the midst of building the project and selling homes, River Run demanded that the city go along with the plans as River Run had already executed them. River Run insisted that the city promise never to allow bicycles on the pedestrian path. River Run advertising featured images of ducks and trout in a "natural" setting, and the market was responding well to the idea of living very close to the river. O'Neill later explained in an interview that he eventually ran out of trust in dealing with the park department. After having sold $80 million of property on the basis that the walking path was on the river and the bike path elsewhere, he wanted some guarantee that the city would not change its mind at a later date and arbitrarily move the bike path back onto the river.[33]

But the Greenbelt Committee was reluctant to place in writing any terminology that "would forever ban bicycling use along the river" and so instructed the park director.[34] In late 1980, after the HCRS finally approved the use conversion and the trade, the Boise city council was free to pass an ordinance accepting the land trade and authorizing the mayor to consummate it.[35] A few weeks later, O'Neill presented the mayor with a document saying that he would consummate the agreement only if the mayor confirmed the Greenbelt path adjacent to River Run would be open only to pedestrians and "shall not be used for the purpose of constructing, developing, maintaining or operating a public bikeway or bike path." The mayor, witnessed by the city clerk, signed.[36]

For the next five years, progress on both paths halted. Despite the protracted and, as Jack Cooper had put it, "excruciating" negotiation, neither party advanced the path for which it was responsible. The park department left the Greenbelt terminus at the entry to River Run. Cooper felt that spending priorities were greater at other parts of the Greenbelt system.[37] Meanwhile, River Run developed several phases of its project, but did not paint the roadway stripes that were to delineate the bike path. The "continuous green belt" ended at the entrance to River Run. By the time progress on the two paths eventually resumed, there had been a flood, new development upstream of River Run, new city policy enacted, more land trades, and new flood-plain engineering—all of which knotted up again and eventually resulted in a pattern of riverfront use quite different from that envisioned by the original Greenbelt planners.

Flood-control engineering had become a dominant influence on the Boise River. The engineers figured out how to redefine the flood plain by building levees and auxiliary channels. The objective of these structures was to remove land from the flood plain so it could be developed with office buildings and houses. With less land in the floodway, the public access to riverfront land was on a much narrower and restricted basis. The auxiliary channel at River Run differed from the one at ParkCenter in that it was a "wet" channel all the time, while ParkCenter's was intended for the conveyance of water only during high water. River Run decided to take advantage of the water by routing some of it through its project as an amenity, designing it in collaboration with the Idaho Fish and Game Department as trout rearing habitat. River Run developers were very proud of this innovation and received desirable publicity and recognition in design awards.[38]

As the engineers narrowed the flood plain, they brought private development closer to the river and at the same time took space away from the still growing public constituency of hikers, bikers, roller-skaters, rafters, and nature lovers. The

**Figure 6.4**　　　　　　　　　　　　　　　　*Susan M. Stacy*
**Through its development River Run threaded artificial streams, an attractive amenity designed to double as trout-rearing habitat.**

numerous bureaucratic oversights, inattention to federal regulations, and failures of coordination among city departments doomed the earlier ideal that the Greenbelt and its paths—including the bike path—not cross vehicular traffic. Neither the Greenbelt committee, the planning commission, nor the city council held a hearing or public discussion on the matter. No one questioned any of the design solutions for flood control created by the developers' engineers, endorsed by the Corps of Engineers, and accepted by the city engineer.

Emkay and River Run developers simply brought their flood-proofing plans to the city as part of the process for approving subdivision plats and proceeded from those. Local and city engineers obtained the blessings for these plans from FEMA, the regulatory agent for the National Flood Insurance Program. No one wanted to violate FEMA standards for fear of threatening Boise's status in the program. Since FEMA standards were less restrictive than those of Boise City's declared "guidelines," and had the engineering expertise of the Corps of Engineers supporting them, they had more credibility with professional engineers than locally generated ideas about using the floodway for public recreation.

A FEMA official approved the Emkay and River Run concept of removing land from the flood plain and floodway at a meeting of the engineers held in April 1977: "Modifications to the floodway and the floodplain would be all right just so other properties were not endangered," he said.[39] That was the operating policy thereafter, regardless of planning policy intentions or adopted city policy. In every case where the city delegated the engineers to solve flood-plain problems, the result was a modified floodway, more flood-plain land for private development, and less land available for the "greenbelt, wildlife habitat, or . . . open space" specified by the general plan.

In 1986, when the public gradually learned that the city's quiet negotiations had resulted in the divided pathway system, land trades that had violated federal rules, delays in extending the paths, and sanctions in the use of further federal funds to help acquire park land, there was an angry response. "I am appalled," wrote one man in a letter to the *Idaho Statesman*, "that our local government has allowed a developer to move into an area on our river and develop homes at a profit, at the expense of our community needs . . . I don't remember at any time that we were asked if it was all right to give up the banks of our river to private use, but this is what has happened at River Run. Our brightest city accent, *the bike path*, has actually been eliminated from the river for several years."[40]

But by this time, the old community consensus that the flood plain was not a suitable place for private development had long been forgotten. Instead, flood-plain land was now regarded by all as a highly feasible location as long as builders observed minimum federal standards for safety. The beleaguered park director had to remind the public that the city did not acquire land or easements along the river because of the goodwill of private landowners, but had to buy it at market rates. The city had to give something in order to get something else. It had to bargain, using land trades if it could, and otherwise scrape for resources in a town that provided too little from the property tax to do everything the public wanted.[41]

## New Metro Plan Policies

However, the storms of 1986 move too far ahead in the story of the evolution of city policy brought about by the National Flood Insurance Program. While the consequences to the Greenbelt of flood-plain planning were remote from public view until 1986, the appearance of the ParkCenter levee was not. Citizens, planners, and others observed the construction of this long levee and worried about its potential impact on downstream or cross-stream properties during a flood. They were at work trying to shape the 1976 comprehensive plan "guidelines" into "policies." During hours of late-night meetings, the developers and their engineers pleaded their point of view. They argued for a policy statement that would allow them to use levees to take land out of the flood plain and floodway if it could comply with FEMA's minimum safety standards. They resented policies that committed to the public any use of privately owned lands in the floodway or flood plain. They felt that such an idea was a value judgment being imposed on them inappropriately by planners and others and were not in the best interests of private property owners.

Developers had several incentives to remove land from the flood plain and to move floodway boundaries close to the river. Partly because of the popularity of the Greenbelt, the land had a high value. Flood insurance was costly for structures in the flood plain. Developers argued that they were already giving something up by not developing in the floodway itself. Structural methods of flood control were the only way they could remove land from these disadvantages.[42] Although the Greenbelt enhanced the value of property adjacent to it, landowners wanted the public to buy any land protected by a setback for public use.

When all the debate was done, the city council at last adopted in October 1978 a new general plan, known as the Metro Plan. The overall goal of the policies was to "prevent costly property damage and threats to life, preserve wildlife habitats, enhance the Boise River greenbelt, and encourage open space/recreation uses." It contained an unequivocal prohibition against future levees: "All development will be floodproofed by means other than levees." Other methods of flood-proofing were unaddressed. Another policy acknowledged the implied message in both the ParkCenter and River Run approvals; namely, that a floodway or flood plain hundreds of feet wide in some areas was too much to ask a landowner not to develop. The Metro Plan scaled down the old commitment of the entire flood plain to the Greenbelt. Instead, open space uses would be restricted only to the floodway and in any flood-plain lands 100 feet from it. Developers could transfer the densities thus given up to portions of their property inland of the 100 foot setback.[43]

The first test of the new 1978 plan policies came two months after they were adopted. Developers of a proposal named Forest River sought to build offices and apartments in the downtown reach of the river between the 8th Street and Americana Bridges across from Ann Morrison Park. Following its usual procedure, the planning commission approved the overall concept plan, but conditioned it by saying that flood-plain arrangements "shall be approved by" the Corps of Engineers before construction could begin.[44] Accordingly, the developer's consultants met with the Corps and received preliminary endorsement of their plan "to develop a dike" [levee]"[45] The levee was relatively low—about four feet high—and designed to meander along the river so that it would be more aesthetically pleasing as part of the Greenbelt already developed along that stretch. The levee would have two feet of freeboard and tie into high ground on either end of the 2,400-foot-long riverside project.

The Forest River levee solution puzzled Ada County's planning director Chris Korte. Landowners outside the city limits in his jurisdiction had been pressing the county for development approvals in the flood plain. The county had earlier agreed to conform to Boise's plan policies in reviewing such proposals. Korte had been recommending denial of any projects proposing to use levees to take land out of the flood plain. Observing Boise's *de facto* approval of levee projects, he asked Boise Mayor Eardley for an explanation. City Engineer Chuck Mickelson responded:

> [I]t seems to me that there are degrees of levee construction . . . it seems to me that early on [in the case of ParkCenter] a commitment was made by the City to approve the development. In order to approve the development certain engineering measures needed to be taken in order to protect [the properties] both upstream and downstream . . . if no levee construction were allowed whatsoever, then development such as ParkCenter and River Run would not have been able to proceed. . . . I believe that the construction of [the Forest River levee] is an excellent choice in that it would remove the U.S. Post Office, the K-Mart site and other areas from the Boise River floodplain . . . the obvious benefit to property owners is that they will not be required to purchase floodplain insurance.[46]

Here was a succinct statement of Boise's economic development priority and engineer-defined solutions to the flood plain. The mayor forwarded the statement on to Korte.

## The Boise River Plan

In 1982-1983, a riverfront development proposal came along that was so disturbing to the city that its string of approvals for development in the flood plain came to an end for awhile. It also set off a new series of public debates about the flood plain and precipitated yet another change to city policy. Roger Crandlemire had a 29-acre parcel on the north side of the river east of town entirely in the flood plain and mostly in the floodway. Trees and thick brush covered the property and one or two old farm buildings stood in clearings on the highest ground. Crandlemire wanted to remake the site to support 70 residential units. He proposed to reduce the floodway substantially, replacing a shallow side stream with a deep dredged channel, filling the low ground, and building levees where necessary to free enough land for the development. He would install a control structure to direct the proper flows into the dredged channel. Someone would have to be responsible for the maintenance and operation of the system in perpetuity.[47]

The Idaho Fish and Game Department entered the picture because the side stream happened to be one of the few places along the river in which trout spawned. Fish and Game officials hoped to prevent the destruction of this habitat. Other allies of wildlife included biologists, planners, and a citizen group that called itself the River Watch Committee, formed to promote the protection of natural and wildlife values on the river. All believed that the development would destroy a significant riparian habitat and the wildlife that relied on it. All made impassioned appearances at the public hearings dealing with the Crandlemire proposal.

Alarmed at an upsurge of sentiment that seemed to endow wildlife habitat with such a high priority, the Greenbelt Committee began to feel that the most basic principle of the Greenbelt—public access to the river—might be compromised. It sent a message to the planning commission asking that if the project be approved, the riverbank should not be removed from the public domain on the premise of designating it as a wildlife preserve. The planning commission heard all of the conflicting testimony and decided to ask the developer for an "environmental assessment" of the site. This was perhaps the first time the commission had ever requested such an assessment and certainly the first time it had ever invoked the Metro Plan goal statement that said the flood plain should be "carefully regulated ... to preserve wildlife habitats."

Corps of Engineers representative Ron Barrett provided two kinds of input to the review process. In a formal letter he replied to the city engineer's request for a technical response to the floodway modification proposal. The levees should be protected from erosion, he wrote, and a "404 Permit" would be needed for the fill work below the high water mark. Adjacent property owners should agree to any changes in the floodway limits that might affect their property.[48] In a second letter to the city engineer, this one more informal, he expressed concern over the

**Figure 6.5**
**A popular site along the Greenbelt near downtown.**

Susan M. Stacy

cumulative effect of the many modifications to the floodway and flood plain that Boise had been permitting. He feared that city ordinances were not adequate to enforce continued maintenance of channel excavations, and that lack of such maintenance could set the scene for disaster. He suggested that the city forbid dredged channels within the floodway when their purpose was to narrow the floodway.[49] The Corps was still sensitive, as it always had been, to the problem of reduced channel capacity in the river and the flood-control benefits of the floodway.

When the Crandlemire team finished its environmental assessment, the city planning staff recommended denial on the basis that all but six acres of the site were wetlands, and that wildlife habitat should be preserved in the flood plain. In the end the city council denied the application after hearing Fish and Game representatives, biologists, and pro-wildlife interests discuss the critical needs of fish, ducks, geese, and bald eagles.

The Crandlemire affair, while resulting in a denial, provoked several uncomfortable questions no one could answer: What would constitute wildlife habitat? What did "preserving" entail exactly? Was all flood-plain habitat equally worth preserving? If it were not, how could you tell the difference? And, above all, what did the city have on its ordinance books aside from the very general goals stated in the comprehensive plan that would protect it from accusations of arbitrariness in approvals and denials?

With the help and advice of the U.S. Fish and Wildlife Service and Idaho Fish and Game Department, the city quickly dispatched a group of consulting biologists to provide the city with a thorough inventory of all the wildlife habitat in the Boise River flood plain. The city asked them to classify habitat areas in relative degrees of importance for preservation. At the same time, the city council deleted the policy in the Metro Plan that had given 100 feet of the flood fringe to open space. The assignment of such a distance seemed somewhat arbitrary in comparison to the new idea that the location of open space should depend on the actual merits of specific habitat areas.

Councilmember Glenn Selander, eager to relieve the study team of any worries about self-censorship in their recommendations, assured the consultants that they were expected to ignore politics and let the council deal with that later, if necessary.[50] After the team submitted its report, the planning department organized the necessary "citizens advisory committee" to evaluate the recommendations and draft new policies on the flood plain and wildlife protection for the comprehensive plan. The proposed amendments soon became known as the Boise River Plan. The committee accepted the biologists' value judgements about the species to be selected for habitat protection. "Species of special interest were selected on the basis of their recreational value, visibility, and/or their unique and sensitive nature," the biologists had written.[51] This translated to four priorities: waterfowl and trout for recreation value; great blue herons because they are "unique and sensitive" and may abandon an area if their nesting sites are disturbed; and the bald eagle, an endangered species also sensitive to disturbance.

The team's report went on to identify the Greenbelt as a negative development for wildlife habitat because the paths replaced trees and other vegetation and because its entire purpose was to attract people, whose presence disturbed nesting ducks and wintering eagles. About a dozen bald eagles wintered along the Boise River; the removal of mature trees and the clearing of brush within 200 feet of the river threatened their continued presence. Areas along the river with the lowest amount of eagle use corresponded to the areas with the highest human use.[52]

The proposed plan stimulated a heated debate among developers, biologists, engineers, and planners over who and what would have proximity to the river. At least one developer, worried about the direction the plan was taking, appealed to Acting District Engineer John Hathaway at Walla Walla, to review Boise's floodplain regulations. The developer hoped that he might possibly find that they were too restrictive and be disposed to intervene on the developers' behalf. Hathaway replied that local restrictions may legitimately be "in excess" of Corps criteria, and that the Corps had no comment on greenbelts, setback lines, or landscape screens.

*Developing the Flood Plain*  99

Further, he agreed with Boise's effort to stop changing the floodway boundaries.[53] Later, Corps analysts reviewed the last draft of the policies and provided comments solely on technical aspects of the language as they related to the Corps' operations and responsibilities.[54]

The discussion wound its way through the advisory committee, the planning commission, and finally arrived before the city council. The draft underwent compromise, refinement, and new ideas at each step. Finally in 1985 the council repealed the old 1978 flood-plain policies and replaced them with the new Boise River Plan. This time the policies prohibited channels as well as levees as methods of flood control and forbade relocating the floodway boundary to accommodate development. The new plan converted the biologists' recommendations into various setbacks and new lists of allowed and forbidden uses in the floodway and other habitat areas. Future developers would have to provide strips of vegetation 25 feet wide along the riverbank to enhance fish habitat. The plan banned the Greenbelt bicycle path from the floodway, forested wetlands, emergent wetlands, and within 200 feet of the 6,500 cfs high water mark in the areas designated as bald eagle winter habitat, the very places where it had always gone before. Pedestrian paths would be allowed, but bike paths would have to be at the edge of such areas or even further removed.[55]

Planners, wildlife biologists, and many other citizens were happy with the new plan. They boasted that "Boise is the only city of its size in the continental United States which has bald eagles wintering within the city limits" and that the new policies were designed to keep it that way.[56] Plan supporters thought the city could now "have it all": a Greenbelt harmonized with wildlife preservation, walking paths located in sensitive areas, and the more heavily used bike paths situated where they would not be a source of disturbance. They hoped the "bald eagle ideal" would enhance the Greenbelt ideal as a mark of Boise's unique character.

## Amending the Plan

But no one could rest on the new plan for long. No sooner was it adopted than its policies were challenged by a new proposal for economic development. It was Emkay again, this time proposing a regional shopping mall at the site of an old gravel pit in the floodway less than a mile west of downtown. A large parking lot would need space where the plan policies called for wildlife habitat. Emkay engineers proposed to redefine the coefficient of roughness to shrink the floodway boundary as much as possible, but it was not enough to provide for the parking lot.

After nearly two decades of struggle, the Boise city council had recently given up its goal to locate a regional shopping center downtown. The effort had failed after years of sustained effort—and after years of sustained voter support of candidates who carried on the fight. The town at last ran out of patience. In the summer of 1985, just a few weeks before it adopted the new flood-plain policies, the city had changed its downtown planning goal and granted zoning for a large shopping mall at a site known as Westpark on the far western edge of the city. Still, the suburban mall had not yet been constructed, and city officials, whose hearts were still with a downtown center, hoped that somehow Emkay could actually bring off a mall project at the riverfront site located so near the edge of the central business district.

The planning staff and Emkay developers searched for loopholes in the Boise River Plan that they had so recently helped to craft, hoping to find some language that could somehow be interpreted to permit the parking lot. They turned up none. The only solution was to have Emkay ask the city to change the plan just enough to permit the project. Emkay asked. Making a narrow change to the river policy seemed to be a reasonable sacrifice to give the Emkay project a chance to succeed. The city said yes and quickly approved the shopping center concept and the parking lot. Despite the city's accommodation, the Emkay venture failed, and the Westpark Shopping Center opened in 1988.

## Economic Development and Garden City

Boise was not the only town interested in economic development in the flood plain. The small town of Garden City, situated on the south side of the Boise River west of Boise, was desperate for new growth. The town had more than its share of transients and poor residents; most of the housing consisted of mobile homes, and the property tax yield was too low to do very much about the town's problems. In the 1970s new municipal leaders determined to make some serious changes. Their strategy was to annex new land and let developers build high-quality residential developments. Garden City would be able to improve its property tax revenues. Taking Boise by surprise, Garden City leapt across the river and began to annex lands on the north bank. The developers kept their promise and built Plantation, a substantial high-quality residential development adjacent to the river.

Not surprisingly, Garden City development standards were easy in comparison to those of Boise or Ada County. The town permitted levees to protect new development, but still wanted to preserve the capacity of the river to carry anticipated flood waters.[57] The Plantation developers, controlling 254 acres, had already come up against Ada County's readiness to enforce its flood-plain policy, so they were more than willing to turn to Garden City for annexation, appreciating the town's more relaxed approach to setting standards.[58]

Plantation did not control all the land immediately adjacent to the river, where the state of Idaho owned strategic parcels. In 1978 the state agreed to swap or lease the land to Plantation on the condition that the developers build and maintain a public bicycle path on the riverbank.[59] This was in recognition both of the success of the Boise Greenbelt and of the state's own interest in a future path connection to a state park further downstream at Eagle Island.

The Corps of Engineers helped Garden City by reviewing Plantation's levee proposals and other development features so the project would comply with the National Flood Insurance Program requirements. Corps officials had reservations about the ability of Garden City to enforce or maintain flood-control arrangements, but the development routine was similar to Boise's. First, the developers decided that their goals were to remove land from its designation as flood plain and have the floodway boundary moved as close to the river as possible. Since the city relied on the review, technical help, and expertise of the Corps, developers sent proposals, maps, and computations off to the Corps. The developer's engineers surveyed new cross sections of the river, so that the Corps could redefine the location of the floodway as precisely as possible. Since earlier Corps work had been conservative, the new maps inevitably resulted in moving the floodway boundary closer to the river.

The Corps evaluated whether the proposal would protect against a 100-year flood without causing a rise of more than one foot in the floodway. The analysts could "laboratory test" a proposal, as it were, using the HEC-II model. If the rise were too great or the levees too low, the developer had to go back to the drawing board. Typically, the Corps suggested better locations for levees and channel improvements. When a plan finally met all of FEMA's standards, the Corps so advised city officials, who could then approve the project and take on the responsibility for monitoring the developer for compliance. In the case of Plantation, the Corps also issued a 404 permit, allowing the developers to place fill in part of the floodway.[60]

Plantation developers removed riverbank vegetation, excavated in the river channel, and denuded and dredged a small side stream in the floodway. After executing the land trade with the state, they showed Garden City revised plans that no longer included a bike path. When state representatives pressed them about this, Plantation developers said that sales were not going well and asked for a time extension to build the path later. "Later" never came because by 1983 foreclosure proceedings were underway against Plantation.[61] In June of that same year the Boise River flooded; the flow of 9,500 cfs was by far the largest release from the dam since Lucky Peak had been built. The dredged side stream and the denuded riverbank at Plantation both eroded, substantially diminishing the back yards of several luxury homes and the potential for a future Greenbelt path adjacent to the river.[62]

Not far downstream, just west of the Glenwood Bridge, another group of investors saw a riverfront development opportunity and also requested annexation to Garden City. The state of Idaho likewise owned land there that the developers

**Figure 6.6** *Susan M. Stacy*
**Garden City officials annexed land in expectations that high-quality housing would bring in much-needed revenues.**

wanted to use. In 1980 the state again permitted the use of its land on condition that the developers install a bike path and greenbelt park along the entire length of the project and dedicate it to Garden City when completed.[63] This project was called Riverside Village.

Developers proceeded to haul fill onto all of the flood fringe up to the edge of the greenbelt path, relieving them and future homeowners of obligations to pay flood insurance. Custom homes were built next to the river on lots for sale at premium prices. Three years later developers built a greenbelt path through the first phase of the project, but then closed it to the public, asserting that houses under construction were being vandalized.

At the same time it closed the path, Riverside Village returned to the Idaho Land Board and asked to change its agreement to move the bike path to a future street. It wanted to create more lots on the river and on the banks of a side stream on the western end of the property. When asked why the bike path could not be placed on a bank of the river or a side stream, the developer's representative, Al Marsden, replied that it would reduce the value of proposed estate lots.[64] The Land Board refused to change the agreement, but neither did it force the reopening of the path

*Developing the Flood Plain*

**Figure 6.7**                      Susan M. Stacy
**Riverside Village developers barred the public from use of the riverfront path after the State of Idaho made public ownership of the path a requirement.**

to the public. The path remained closed to the public through 1985 and later. Advertisements for Riverside Village in the *Idaho Statesman* later referred to "the serenity of evening walks on a *private* greenbelt along the banks of beautiful Boise River."65

## The Bald Eagle Ideal

With a slowdown in the region's growth rate in the early 1980s (coincident with a national recession), the third post-Lucky Peak decade came to a close. The Boise River and its Greenbelt were the pride of the community. The decade had brought two revolutions to the Boise area. Implementation of the National Flood Insurance Program had helped to bring about an overwhelming reversal of the idea that the flood plain was useless for anything but outdoor recreation. The appeal of the Greenbelt park stretching east and west of town had beckoned developers and homeowners, who did what they could—and paid what they could—to get as close as possible to the now-lovely stream.

The other revolution was written into local planning policies. As the early developments like ParkCenter and River Run matured into prestigious addresses behind their levees, the public increasingly wanted to preserve the natural qualities of the river, and even to restore natural qualities where they had long been degraded. Policy makers started out in the early 1970s by dedicating the flood plain to open space. Then, they divided the flood plain into a floodway and a fringe and said, "Leave the floodway to open space." But they had left room for the floodway to be narrowed. Seeing their first levees, they said, "No more levees." After they had seen auxiliary channels, it was, "No more channels." Finally, there didn't seem to be enough room for all the activity they would have liked in the Greenbelt, so in the interest of wildlife habitat preservation, they decided to restrict their own access to the river, "Pedestrian paths in the Greenbelt, yes, but bicycles, no."

The progress of public sentiment contributed inadvertently to the interests of those wishing to develop and live along the river. During the frenetic growth period of the 1970s, developers consistently resisted setback regulations that would distance their building lots from the river. They wanted to market riverfront access to their clients and tenants. City officials, forced to decide between open space in the flood plain or economic development, usually chose economic development, progressively narrowing the open space remaining for public use. The mayor of Garden City, Margaret Mockwitz, wrote to the Boise mayor after the habitat survey had started, saying that Garden City was finally getting some high-quality urban development. "I will resist any attempt by special interests to lock up the Boise River in the interests of wildlife only."[66] Residential promoters gave their potential customers as strong an impression as they could of exclusive access to the river. Advertising brochures, full of photographs and evocative drawings, portrayed ducks floating serenely on quiet ponds, the reflection of vegetation dipping onto the river, and an occasional couple or individual strolling or fishing. "Watch the wild water fowl and enjoy the grassy open spaces... a beautiful natural setting reflecting your success," said a Riverside Village brochure. "The Boise River flows slowly next to our shade covered paths."[67] For the Island at River Run, the brochure said, "Light and shadow, murmuring waters, plash of trout and play of wind, bird-song and rose-scent," and pitched wooded acres, serenity, privacy, seclusion, and security.

Such development values became less and less compatible with the early Greenbelt ideal of "perpetual unrestricted access" by the public, especially when the public had the mobility provided by bicycles. Despite the public's growing embrace of habitat preservation values, there were now several stretches of the river where the enjoyment of those values was reserved to the few, and where the public was excluded or discouraged.

The Corps of Engineers played its part in the development of the Boise River flood plain according to its traditional mission—national economic development with the help of engineered solutions to bring about safety from flood hazard. As

*Developing the Flood Plain* 105

usual, when there were local conflicts the Corps relied on its engineering skills, declining to intervene. Local authorities (and local engineers for whom flood-plain engineering was not a specialty) regarded the Corps as the best possible source of technical assistance for plans to remove land parcels from flood-plain designations and to install flood-control structures. With Corps input, officials permitted levees, diversion works, channel dredging, and auxiliary channels. The Corps' support of economic development in the city continued the work of Lucky Peak, which had supported economic development in the desert.

# Chapter Seven
# The Flood of 1983

After 30 years it was obvious that Lucky Peak reduced the annual flood risk on the Boise. The U.S. Army Corps of Engineers had managed its operations so that the highest spring flow had never exceeded 7,500 cfs. In most years the Corps held it to 6,500 cfs or less, a level that admittedly caused some damage when that flow remained constant for days or weeks at a time.[1] Nevertheless, a flood like the one in 1943 became a faded memory.

Lucky Peak had changed the river and life along its banks. The absence of the old-time flood crests changed the appearance of the river. More islands lay in the stream. Vegetation anchored the islands and grew to maturity. The citizen-sponsored interest in parks and recreation brought about the Greenbelt, and the Greenbelt led to the impressive office parks and subdivisions in the flood plain. The changes reflected the work of many policies, many agencies, and many shifts in the attitude and consciousness of the public. The National Flood Insurance Program, the environmental movement, the Corps of Engineers expertise, the 1970s growth boom, the Boise River Plan—all played a part in liberating the flood plain from its earlier days as a wild and wasted place. The river was the scene of lively new and mostly successful economic development. Few people missed the gravel extraction industry or the food processing plants. Newer citizens in town, who never had known the river as the dump it once had been, sometimes used the term "pristine" to describe the vegetation along the river.

Everything was in place for a good test of the flood-control system. The new and improved rule curves were in use, homes and office buildings rested comfortably behind their low-rise levees, and various headworks and capacity-making channels awaited their moment of truth. In 1983 the moment came.

"The Boise River is not expected to flood," said the Bureau of Reclamation river manager Harold Brush in late May that year. However, the headline on the news article that quoted him warned that "High Water Poses Threat to Idaho Fun-

Seekers."[2] The mountains behind Boise were loaded with snow, much of it having fallen earlier in April and May. The "fun-seekers" were responding to exceptionally warm weather—94 degrees on May 27, which set a new record for the date—and were evacuating Boise for a long Memorial Day weekend of camping and fishing. Seven dirt roads in the Boise National Forest already had been washed out by creeks swollen with melted snow. Authorities cautioned campers that the roads they crossed going in to their camp sites could very well wash out behind them. The *Statesman* was full of other news about high water in the region, the unusually high May snowfall, a bridge that had washed out on the Boise's South Fork. At that moment the Corps was releasing 6,500 cfs from Lucky Peak and thought there was plenty of storage space left in the Boise River reservoir system.

Memorial Day was a scorcher. Temperatures persisted in the low 90 degrees. The National Weather Service placed a flood watch on the Snake, Big Wood, Little Wood, Payette, and Salmon rivers—drainages all around the Boise. Reservoirs on those rivers were filling fast. Meanwhile thundershowers were predicted for the mountains above Boise.[3] No flood watch was mentioned for the Boise River. The last day of May brought news of flooding and mud slides at Salt Lake City in Utah. It was still hot. Responding to an inflow of snowmelt estimated at 26,000 cfs, Lucky Peak operators raised the Boise River flow to 7,000 cfs. On June 1 they raised it another 500 cfs and declared that the river would go no higher.[4]

But it did. When flows reached 7,500 cfs, a coming flood still unannounced, the public began to react. Ada County Civil Defense Coordinator Jack Blake contacted individual families he knew would be threatened at well-known low spots. Property owners started their own independent sandbag projects. Ada County sandbagged Barber Park to protect one of its access roads. The Boise Park Department

**Figure 7.1** *Boise State University Library*
**The 1983 flood approaches a roadway in Barber Park.**

sandbagged parts of the Greenbelt and watched other parts go completely under.[5] Cottonwood trees began to topple into the river, the soil around their roots eroded away. Some of them fell in such a way as to impede the flow of the river or to divert water closer to or over a levee.[6]

On June 6 the Corps District Engineer Colonel Robert Williams from Walla Walla met with hydrology and disaster services personnel in Boise and told the press, "We can not afford uncontrolled flows on the Boise River. That is our first priority. We're playing this an hour at a time. We will only have half a day at most."[7] This ambiguous message referred to the Corps' intention not to let water flow over the Lucky Peak spillway, but rather to send water through the outlet tunnel to control releases as needed. Corps officials planned to issue warnings prior to each 500 cfs jump in flow. The snow runoff was flowing into the reservoirs much faster than water was being released. They were filling rapidly.[8]

Corps operators decided to ratchet the river 500 cfs at a time to help property owners determine how much land would go under with the next jump, giving them time to prepare accordingly. Each 1,000 cfs was expected to cause a rise of four to six inches in the water level. In Canyon County, 7,500 cfs put 90 acres of farmland under water. At 8,500, it was expected to inundate 200 acres.[9] Fortunately, nothing interfered with this plan: a break in the New York Canal would have required sending the 2,700 cfs canal release down the river or holding it in the reservoir, accelerating its fill rate. In that event, citizens would have had to react to suddenly increased flows and rely on the state emergency plan, as well as themselves, for help.[10]

Property owners and local government officials responded to the rising river in several ways. Joan Carscaddon, a homeowner on Eagle Island, moved her things out of her home as the flood washed out a levee and the road to her house. Her neighbors commented that in the 14 years they had lived there the flooding on Channel Road had gradually grown worse due to the growth of trees in the Boise River channel.[11] Other farmers and homeowners in low-lying places evacuated animals to higher ground and built levees around their homes.

The buildings at ParkCenter, Riverside Village, and other flood-plain developments had been permitted with a number of conditions assuring that neither they nor other properties would be damaged by a 100-year flood. Developers had promised as a condition of approval that flood waters would flow unimpeded in the designated floodway.[12] The HEC-II computer model of flood behavior could only predict reliably if the flood were given the space the floodway provided for it. Otherwise water might go where it was not expected—onto property not previously in the path of the flood, not prepared for flooding, and not protected by insurance. But property owners and officials forgot these promises as they watched the advancing flood eat away at their fairways, paths, and landscaping.

*109*

**Figure 7.2**  *U.S. Army Corps of Engineers*
**Aerial view of cottonwood trees that fell into the river when the flood eroded the banks of the river.**

*Boise State University Library*
**Figure 7.3**
**Part of Barber Park is inundated by the flood.**

**Figure 7.4**  Susan M. Stacy
**ParkCenter Boulevard lay in the path of the engineered auxiliary floodway. Boise River is to right. Water on road was caused by seepage.**

At ParkCenter, the river threatened to flow over the low spot in the auxiliary channel levee. Alarmed executives from nearby buildings sandbagged the entrance to the channel. Charles Winder, a former Emkay representative who was a party to the agreements regarding the system, phoned City Engineer Chuck Mickelson and argued that Emkay should be able to block the channel because the city had restricted the floodway across the river by sandbagging the golf course. Emkay felt that this was causing the river to rise faster than it otherwise would have. An hour after Winder's call, a FEMA official, alerted by someone from Boise, called Mickelson and said, "What's going on? We understand they're putting sandbags on the dike. That's the last thing that should be happening up there." Mickelson ordered the sandbags removed. Employees at the office building built a three-foot sandbag wall to protect transformers and other equipment. The corporate executives were distressed, understandably feeling that the flood belonged in the river. Later one of them told the civil defense coordinator that he had not expected the city to enforce the auxiliary floodway.[13] But they left the channel alone after that.

Four of the fairways at the Municipal Golf Course were partially covered with standing water. Golfers brought tractors and sandbags to build a restraining line at the edge of the course, well within the floodway.[14] Whereas city officials had kept open the auxiliary floodway at ParkCenter, they allowed the sandbag levee on the floodway portion of the golf course to remain. Although the principle of permitting

The Flood of 1983

**Figure 7.5**  *U.S. Army Corps of Engineers*
Sandbags at the edge of Municipal Golf Course helped hold the flood to standing water on the fairways.

the river to flow unimpeded in the floodway might better have been exercised consistently, not all properties had been developed with specific conditions on their land-use permits.

At Riverside Village in adjacent Garden City, where a dredged lake had been permitted and designed to function as part of the floodway, the developer, with help from Garden City Public Works Director Charles Smith, sandbagged against the river. "Why would we destroy $10,000 worth of bike path when two rows of sandbags would do the job?" said Smith.[15]

As the river went to 8,000 cfs, some houses flooded and the river flowed through one of the barns at the fairgrounds. Most of the agricultural damage to crops, levees, and irrigation structures occurred downstream of Eagle Island. The river chewed up the riverfront yards of the homes at Plantation. Lee Krogh of the National Weather Service observed, "The river has silted up and these are trees in the [old pre-Lucky Peak] channel." Anderson Ranch Reservoir filled up, Arrowrock was 97 percent full, and Lucky Peak was 78 percent full. It was filling at a rate of 8,000 acre feet per day and had 57,000 acre feet of space remaining in the reservoir. Corps public information official O.C. Duggar said that if Lucky Peak were full, Boise would have already experienced $65 million in damages.[16]

But still, some space remained available in Lucky Peak. The weather stayed unseasonably warm, and the snow line was only at 7,100 feet. One fifth of the season's snowpack still lay above it.[17] On Thursday, June 9, the National Weather Service predicted that if the temperatures remained high, Lucky Peak would be completely filled by the following Monday. The Corps predicted that there would

**Figure 7.6** *Boise State University Library*
**Volunteers at Riverside Village place sandbags, restricting flow of river in floodway.**

be $1.5 million in damages if the river went to 9,000 cfs and $2.7 million if it went to 10,000 cfs.[18] Arrowrock was completely full, water plunging over its spillway. Lucky Peak had 43,000 acre-feet of space left. Aircraft took off for observation flights over the Boise River watershed. Observers reported grimly that there was more snow than previously had been thought.[19]

On Friday the weather moderated, but the system was still in crisis with runoff exceeding releases. The Corps let go 9,000 cfs on Saturday and 9,500 on Sunday. Hundreds of trees lay across the river by now, creating dams behind bridges and diverting water onto levees. Canyon County farmers had learned by experience that it took 24 hours for a new crest to reach them after the Corps raised the flow. They shored up their dikes. A number of families who lived in mobile homes near the river abandoned them. Not everyone who wanted to buy sandbags could get them. A news reporter interviewed one bagmaker in Boise and learned that he had shipped 67,000 bags, his entire stock, to Salt Lake City a few days earlier and had made 23,000 more for Boise, but was now out of them. "We weren't ready for this," one employee said.[20]

The moderate weather held, and the great snowpack melt slowed considerably. On June 12 the rate of flow entering the reservoirs was down to 16,500 cfs. River operators eagerly waited for it to diminish to about 13,000 cfs, the flow that would equalize the combined flow they had been releasing to the river and to the irrigation canals.[21] As the next week progressed, the Corps continued the flow at 9,500 cfs to draft the reservoir, a precaution against another episode of high temperatures. Fortunately, the danger passed. After all of the information had been assembled and analyzed, it became clear that the peak runoff had been 24,294 cfs, higher than that

*The Flood of 1983*

**Figure 7.7**  *U.S. Army Corps of Engineers*
**General view of floodwaters downstream from Boise.**

of the 1943 flood. The flood flow of 1983 was named a 50-year flood, having a two percent chance of occurring in any given year.[22]

The Boise River flood-control system indeed had been tested. It had prevented a repeat of the 1943 flood and prevented 1983 levels of damage and disruption. Nevertheless, all parts of the system had not entirely passed the test. In addition, the aftermath of the flood left many questions unanswered. The chief one: what would have happened if the weather had continued in the 90s for two more days?

"Sandbags belie floodway promises," said a *Statesman* editorial. Here was one part of the flood-control system that had not passed the test. The real problem, declared the editor, was relying on developers to keep promises 50 years from now when those promises are not so fresh in anyone's memory. The *Statesman* advocated that all jurisdictions stop letting developers reclaim floodway land by making channels to divert water in a flood. The editor made a tacit distinction between developers who had been given permits on specific conditions and all of the sandbagged properties where use of the flood plain pre-dated the system of flood-plain permits. About the latter, there was no comment.[23] Thus, property owners and public officials of Boise, Garden City, and Ada County who had tolerated, abetted, or sponsored their own in-floodway sandbags received no criticism.

The variety of commentary and justification after the flood reveals the gulf that can exist between a computer model of how a system should or will work and how human beings actually behave in the crisis of managing a flood. Chuck Steele, a FEMA official, reminded the community that 702 property owners in Boise and Garden City were able to get flood insurance because the cities had promised in good faith to comply with the system requirements. Ron Barrett of the Corps said,

"I wish no one would change the flood limit lines. That's my opinion, not necessarily the Corps.'" Jeff Youtz, the president of the Garden City Council, had not been on the council when the Riverside Village agreements were signed and said he was not aware of their provisions. Garden City Mayor Margaret Mockwitz said, "We had a lot of fine people out here trying to save some property and I think everybody should receive a pat on the back." The Idaho Water Resources Board was aware of the sandbagging done at Garden City but did not stop it because it did not seem to be posing a hazard across the river. "We are concerned about reasonableness," said Water Resources representative Ervin Ballou. Ron Barrett agreed with the *Statesman* editor when he observed that if a plan could not be relied on three years after an agreement was signed, it probably would be violated again.[24]

The art of predicting and estimating flood damage had advanced little since 1943. In 1974 the Corps of Engineers had predicted that a flood of 10,000 cfs would cause $540,000 of damage in Ada County. In the midst of the 1983 crisis, the Corps spokesperson had said that if the river went up to 9,000 cfs, the damage would be at $1.5 million. As 9,000 cfs arrived and then was hiked up to 9,500, the $1.5 million in damages did not materialize. It was hard to tell whether another 500 cfs would have caused another $1.2 million in damage, as the same Corps spokesperson had said it would. Damage was so slight after the 1983 flood that the Corps did not even conduct a flood-damage survey because there was not enough structural damage to justify one.

Figure 7.8
Susan M. Stacy
**Flood waters fill bike path tunnel under north bank at Main Street Bridge.**

*The Flood of 1983*

With no government agency sufficiently interested in the actual damages or costs caused by the 1983 flood, it is difficult to estimate them. Civil Defense Coordinator Jack Blake compiled the only available data. His total was $146,900. It included estimates of $20,000 for personal property damage, $45,000 for government employee overtime, and $51,000 for erosion repair and other property protection work.[25] His estimate did not include any repair expenditures made by farmers, homeowners, or public agencies in the months after the flood.[26]

Certainly the public sustained losses in the Greenbelt and riverside parks not included in Jack Blake's list of damages. Many of the hundreds of trees that had fallen into the stream were dragged out of the river at public expense. In August, after the most hazardous snags had been removed, Boise Park Director Jack Cooper estimated that 140 still remained as dangers to river rafters and tubers.[27] Municipal Park, directly opposite ParkCenter and at the outside edge of a bend in the river, lost a chunk of land 300 feet long and 55 feet deep in parts. Cooper noted that the lost land appeared to be natural soil, with no sign of the debris, car bodies, and other fill material found in other places along the river. The soil became saturated and broke away when the flood began to subside. Cooper ordered contractors to dump

**Figure 7.9**
*Susan M. Stacy*
**The Park Department dumped riprap along the bank across from ParkCenter to prevent further erosion. This view was taken later in the summer.**

boulders and concrete to protect the remaining edge, but it was too late for the bank, the sprinkler system, the trees, and the bicycle path.[28]

Observers wondered whether the sandbags and permanent levees upstream and across stream had contributed to the erosion of Municipal Park by increasing the velocity of the flood, but no one attempted to analyze this question seriously.[29] It remains for a property owner damaged in a future flood to discover whether unplanned rises in flood waters caused by floodway constrictions upstream result in accelerated erosion or other damage, and if so, who is liable for the damage.[30]

In contrasting the public management of the 1943 and 1983 floods, the system in 1983 might also be found wanting. In 1983, flood news and instructions were largely left to the press to interpret. In 1943, such news had been issued regularly and directly to the public every day at lunch time by the authorities themselves. In 1983 the public perception of a flood hazard materialized slowly. After the flood danger had passed, the *Statesman* reporter responsible for covering the flood complained that he could not get consistent estimates of the danger to the community from the authorities. They did not all agree on the seriousness of the forecast.[31] Aside from indicating that the emergency was under less than unified management, the remark reveals a difficult dilemma for river managers. When officials make hour-by-hour responses to the indicators, they cannot equip the public adequately to organize and plan for disaster. On the other hand, an alarm that turns out to be false might arouse public hostility and resistance to the next alarm.

**Figure 7.10**     *U.S. Army Corps of Engineers*
**The 1983 flood defeated early efforts to keep the river off a warehouse site in Garden City.**

# The Flood of 1983

The 1983 flood was not caused by what might in other kinds of disasters be referred to as "human error." Rather it was a reminder to the community that natural forces could still confound the highly controlled river system. Equipped as it is with rule curves, snow surveys, computerized satellite information transmission, and highly skilled specialists at the controls, the system still has a natural vulnerability: abundant snow or rain in April, May, and June. In 1983 the moisture arrived in May long after the river operators had expected to "make the space" for the runoff. The reservoirs were already nearly full. And the moisture forecast turned out to be 10 percent too low.[32]

The 100-year flood on the Boise is defined as a flow of 16,600 cfs or more. This flood will possibly occur when conditions are similar to those of 1983. Warm temperatures will melt the late season snow so fast that it simply fills and overtakes all the reservoirs. The total flow would most likely be considerably less than the 100-year flood of the *un*regulated river. The levee and other structures below Lucky Peak Dam, on the other hand, will have to do an excellent job of controlling the flood, because homes and offices are now located in the former flood plain. The Boise River reservoir system has a dual mission to provide irrigation water and flood control. The levee and channel systems protecting buildings in the flood plain are designed for the 100-year flood. The total system provides less defense against floods greater than the 100-year event.

**Figure 7.11** *Susan M. Stacy*
River Run levee and channel system secures homes behind them from the flood, but water inundates publicly owned islands.

In summary, the 1983 test of the reservoir-levee system of flood control on the Boise River demonstrated the following: that public notices for crisis planning were shorter than they were in 1943, that the three-reservoir control system could allow a flood twice the volume of that in 1983, that property owners and public officials in Boise did not necessarily react according to approved permits and policies, that erosion protection for properties opposite leveed banks was not planned, and that there was no program for keeping property owners or public authorities current regarding the channel capacity of the river or changes to it.

Although Lucky Peak Dam had not been built because of an emergency need to provide flood protection for Boise, it is certain that Lucky Peak is necessary for such protection today. The flood plain is studded with commercial and residential buildings worth millions of dollars.[33] With their first floors one foot above the 100-year flood elevation, the buildings were safe behind levees and special channels in the event of a 50-year flood—as of 1983.

# Chapter Eight
# When the River Rises

The Greek historian Thucydides once described a plague epidemic in Athens, detailing its terrible symptoms and the treatments given. He wanted medical experts in the future to be able to recognize and treat other outbreaks in the city, should they occur. Ever since Thucydides' example, historians have hoped that their histories might serve the public purpose of improving the future via knowledge of the past.[1] The history of flood-control policy on the Boise River cannot (yet) be compared to a plague—or even a Greek tragedy—but it does suggest ideas and trends worthy of reflection in considering future action.

## Channel Capacity

The most obvious development on the Boise River itself is that the capacity of the channel to carry a flood has diminished over time. Siltation of the channel has been a notorious problem of long recognition, first attributed to mining and timber practices high in the watershed. Then irrigation diversion structures in the valley helped to slow the passage of water, allowing sediments to drop to the bottom. When Arrowrock and then Lucky Peak began to restrain spring floods, they reduced the scouring impact of floods, which had acted to move accumulated sediments out of the valley. Even though the dams act as silt collectors, sediments enter the river below them from agricultural return flows, urban runoff, and tributary creeks north of the city.

An increasing use of flood-plain land accompanied the diminishing channel and, consequently, the river was the focus of an ever-intensifying water-management program. Besides the dams, the community built diversions, levees, revetments, and channels and filled, dredged, riprapped, and straightened the river. Future policymakers are likely to choose from the same array of responses to cope with a continuing reduction in the river's flood-containing competence. They can build more and higher levees, dredge the main channel, create new side channels, or even build another dam. In the past, such responses often have been made in the wake of a flood. Another damaging flood will possibly call forth similar initiatives.

"Conditions in the river change all the time. Trees and bank vegetation change the behavior of water, river gravels move, the channel gets silted," said Chuck Mickelson, Boise city engineer, in an interview.[2] Much of the flood-control engineering that took place in the seven years prior to the 1983 flood assumed a somewhat static condition in the river. Flood levees were designed on the basis of a profile of the Boise River channel unlikely to remain the same over a longer period of time. Yet there are no programs in place at this time to provide a regular review or survey of changed conditions.

The question is whether the next 50- or 100-year flood will occur in a river with channel characteristics substantially similar or substantially different than those that existed when the present flood-control structures were built. Past experience suggests that the river will continue to accumulate silt. The 1985 Boise River Plan proposed one feasible siltation-removal technique as a goal—an annual "flushing flow." This technique would involve sending a high surge of water down the river in the spring for a short period of time to imitate a flood. However, the city has been unable to implement this policy because it does not control the river; and the U.S. Army Corps of Engineers would not consider releasing "flushing flows" without analyzing costs and benefits in comparison with current practices.[3]

It might be argued that there were margins of safety in the design of flood works that would delay the negative effects of a diminishing channel capacity. Conservative coefficients of roughness in defining the floodway and requirements for extra levee freeboard might be examples. But developers negotiated, sometimes successfully, for reductions in levee freeboard or adjustments in roughness coefficients based on a less conservative (and arguably more accurate) appraisal of ground surfaces. No one involved in these activities will say that safety was compromised, but surely the margin of safety was reduced.

Figure 8.1  *Susan M. Stacy*
After 1985, residential development along the river continued to appeal to persons desiring exclusivity and environmental values.

## Diffusion of Management Responsibility

Another trend visible over time is that responsibility for policy on the Boise River has become more and more diffuse. Numerous authorities at every level of government claim one or more specific roles in river management and deny responsibility for other roles. Some agencies claim a shared role, like the Corps and Bureau in predicting the snow pack. In the 19th-century floods, property owners in the flood plain were on their own, responsible for their own decisions and their own protection. Gradually, local authorities (such as the county commissioners who decided that the Broadway Bridge had to be protected at the possible expense of a group of houses) took over control decisions during a flood. The city of Boise undertook its program of lowland filling and channel straightening in order to protect such public uses as bridges and the municipal airport (which occupied the present site of the campus of Boise State University). Then came federal sponsorship of the dams and the management of the river releases at each dam for irrigation, flood control, and recreation.

After Boise joined the National Flood Insurance Program, responsibility became even more diffuse with the creation of a federal agency with a new mission—the Federal Emergency Management Agency. FEMA's mission is to keep federal flood emergency costs down and to establish the regulatory framework for the National Flood Insurance Program. The Corps of Engineers is FEMA's engineering expert in water resources but also provides the same expertise to help communities build structures within the flood plain. The Corps has no regulatory control over land uses in a flood plain, although it does issue permits to place structures in a navigable waterway and "404" fill permits. At the state level, the state issues permits for the "alteration" of a stream. Local communities adopt local plans but use a combination of federal minimum standards and local discretionary standards in their ordinances. Developers are obliged to construct approved plans, but federal authorities who review them on paper do not necessarily get an opportunity to monitor all of them in the field. Local governments, particularly small ones, do not always have enough of their own expertise to do that monitoring either. In a flood, (no matter what river it may be) people protect their own property, unaware of the impacts downstream, and certainly unaware of the invisible and very abstract boundary between floodway and floodway fringe. And often, as the 1983 flood event in Boise demonstrated, public officials who do understand the regulatory systems desire to help the public protect that property. If a flood were to expand its territory downstream and do unexpected damage, who is responsible? Is anyone liable?

As injured parties more and more go to court to find answers to these kinds of questions, the Boise River case shows why a judge or jury might have a difficult time deciding the question. The following pair of informal comments from individuals involved for many years with flood control on the Boise highlight the problem. The first reflects the perception of one Corps of Engineers employee responsible for managing the flows from the reservoir during the flood season:

> No, *we* [the Corps] are not helping people build in the floodway. It's the city of Boise who is responsible. Development means revenue. The city should have made a greenbelt in the flood plain.[4]

The second comes from the Boise city engineer, who felt comfortable with the construction in the flood plain because the expertise of the Corps was backing him:

> We use the Corps as our experts. If the Corps is satisfied [with revisions to floodway and flood-plain calculations], if the Corps feels it's a prudent approach to take, we have accepted that. They are our technical consultants. I think they have been reasonable.[5]

To a flood-damaged litigant, the comments might seem as though the city and the Corps were mutually denying responsibility for the course they followed. Taken together, the comments reiterate local commitment to economic development in the flood plain and a perhaps reluctant but *de facto* federal acquiescence.

## Ultimate Responsibility?

If a blurring of responsibility on the river is an outcome of its history, then what about the original decisions that led to that outcome? Where and what was the motivation for decision-making? Various studies of the Corps of Engineers and its projects have examined three possible sources. Decision models have focused respectively on federal bureaucracies, Congress, or local interests.

For example, Arthur Maass focused on the responsibility of the Corps of Engineers as a public service agency. In his 1951 critique of the Corps, *Muddy Waters*, he wrote that an administrative agency "should be responsible for formulating as well as executing public policy."[6] Policy is formed in the very exercise of discretion in everyday operations. Administrators are in a position to propose policy because they have an opportunity "to observe at first hand how policies work out in practice . . . have a shrewd appreciation of what is practical and what is not," and have an "ability to represent interests which are not well represented by organized pressure groups, for example the consumers."[7] By Maass' model, the Corps might have taken into account the public interest of the Greenbelt users before it helped Boise reduce its width. Greenbelt consumers are a typical example of a group that is not well organized.

His standard seems to burden the agency without giving due regard to the role of other participants in the decision stream. Maass based much of his criticism of the Corps on the story of the Kings River project in California's Central Basin.

There the Corps had commandeered a project that had more of an irrigation justification than flood control. He portrayed the Corps as an irresponsible force bent on exercising its will alone. Only briefly did he mention the supporting role played by "local interests" who backed the Corps and had much to gain if the Corps succeeded in exercising its will—namely, a dam built at full federal expense. Maass spared criticism of local interests, thereby discounting the significance of local power. But the Corps, he asserted, "fails to live up to a great many of the accepted standards of professional responsibility."[8]

Other writers have emphasized the responsibility of Congress for approving flood-control projects with relatively weak flood-control benefits. John Ferejohn, in *Pork Barrel Politics,* illustrated the point with the 1944 Roanoke River Basin Project. Roanoke's economists had to add hydroelectric or irrigation benefits to obtain a favorable benefit/cost ratio. Ferejohn could easily have made the case with Lucky Peak—or any number of others.

But Congress had broader concerns at heart at the end of World War II. It seems clear that multipurpose reservoirs like Lucky Peak were valued as stimulants to economic production and growth and as preventives against another depression, an attitude shared by local interests who wanted a share of national wealth and prosperity.[9] Ferejohn blamed a log-rolling Congress, or more particularly "the structure of Congress," for accepting weak benefit/cost projects.[10]

Finally there are decision models that give more weight to local interests, especially in post-war projects. And these are the models the Boise River case supports. Economist Leonard Shabman compared the way water projects were *actually* initiated with the *ideal* way they were supposed to be initiated. The ideal was that "comprehensive river basin planning" for a stream or river system would generate candidate projects; to these a benefit/cost analysis would be applied, and only those with greater benefits would be built. But Shabman's analysis demonstrated that, invariably, local interests who had a water problem initiated projects. Decisions about whether to proceed with a project were made early, using what he called "engineering and political judgment." After that, benefit/cost analysis simply legitimized the decision.[11] It would be hard to find a better example than Lucky Peak to ratify this model.

While critical evaluation of the Corps—and of Congress—certainly has its place, it is also important to recognize the heavy weight of local responsibility in policy setting. In Boise, during the 1940s phase of structural flood control, local responsibility was paramount. Economic development pressures and appeals for flood relief clearly sprang from local interests. It was those interests who recognized the 1943 flood as an opportunity to promote a new water-conserving project, who coordinated their efforts with the governor, who coached the congressional delegations, and who created and financed local "promotional" organizations.

During the flood-plain development period of the 1960s and 1970s, it took advisory committees, planning commissions, administrators from many government organizations, the governor, county commissioners, city councils and mayors, and alert (or somnolent) citizens to account for the outcome of decisions regarding water resources and flood-plain management on the Boise. The politics were unequivocally local.

If local interests supplied the motive power for decision making, then what role did the Corps play, if any, in influencing the decisions? As noted earlier, during the 1960s and 1970s, national representatives of the Corps pointed proudly to its handful of nonstructural solutions to flood problems, among them the use of the flood plain for recreation or open space. In Boise the Greenbelt plan was in place as part of a municipal improvement project before the pressure to put buildings in the flood plain emerged. When local conflict ensued, the Corps avoided taking sides. The Corps watched the city reduce its flood-plain setbacks, approve levees, and permit development, albeit in accordance with the minimum standards of the NFIP, in the flood plain. Corps employees like Ron Barrett obviously had opinions critical of this, but they conscientiously stayed out of "local politics," offering only technical input, officially silent about the wisdom of the city's direction. Perhaps the Corps had an opportunity to lead, as its national representatives seemed to be doing, but regional district employees regarded local conflict as a constraint, not an opportunity. But for that, Boise might have been one of the places in the country warmed by the glow of a "new" environmentally sensitive Corps of Engineers.

Of further interest in connection with the Corps of Engineers is its response to the National Environmental Policy Act (NEPA) and its requirement that environmental impact statements (EISs) evaluate proposed federal actions. Students of the Corps have wondered how NEPA changed the Corps, how environmental litigation forced the Corps to change, or how well NEPA improved the quality of decisions made because of the required environmental impact statements.[12] The Walla Walla District performed an EIS on Lucky Peak Dam, even though it was a *fait accompli*, built long before the law required EISs. Although some Walla Walla Corps employees thought the requirement to perform EISs on completed projects was ridiculous, others thought that a review of the environmental impact of operations was a positive thing.[13]

When the 1974 *Draft Environmental Impact Statement* for Lucky Peak failed to propose a solution to minimum stream flow needed for sewage treatment plant effluent dilution, the angry reaction from local municipal interests was equal in impact to environmental litigation in other parts of the country. It brought about change, even though the key issue was the protection of taxpayers rather than some natural feature of the environment. Boise used the EIS process to enhance municipal fiscal goals at the expense, some would say, of environmental quality, a somewhat unusual use of the environmental impact statement process.

## Urbanization of the Flood plain

Another pattern in the evolution of the Boise River flood plain pertains to the phenomenon of its urbanization. Urbanization was both preceded by and stimulated by federal flood-control structures. Lucky Peak Dam did not derive its constituency from among urban investors or property owners. Riverfront industries such as shipping and warehousing were never factors in Boise's growth—as they might have been in other American cities susceptible to flood hazard. The agricultural products of the valley were shipped overland or by rail. Use of floodplain land, even in the city, was retained for farming, public uses, and gravel extraction. Food processors used the river as a depository for waste. Urban land uses up to 1943 had not become valuable enough to justify the flood protection they obtained from Lucky Peak, although Corps studies did portray urban encroachment into the flood plain as "inevitable."

Despite the increased protection from flooding introduced by Lucky Peak Dam, the most immediate stimulant for constructing homes and office buildings in the flood plain was the shift in values brought about by civic pride and the environmental consciousness of the 1960s and 1970s. After the grass-roots public started cleaning up the river and investing community energy and funds into a riverfront park environment, private investors decided the results were attractive enough to risk private capital adjacent to it. Developers willingly bore the extra expense of reclaiming gravel pits and building flood-control works because the returns promised to justify it. To assist in the effort, federal agencies lent their engineering expertise and provided a flood insurance program.

## Valuing Public Goods

The Boise River illustrates the difficulty that public policy analysts and economists traditionally have had in assigning a dollar value—or other appropriate value—to "public goods," resources that are not privately owned, but are shared in common by all. For example, in 1950 the Department of Agriculture reported on the impact that an improved or restored watershed would have on reducing flood hazard in the Boise River watershed.[14] The value was hard to quantify and hard to relate to economic development. Relatively little effort has been made in this area of flood hazard reduction. Bridges supply another example. Erecting bridges that would have survived floods would have saved the public from fairly expensive flood damage, but no benefit/cost analysis seems ever to have been performed on that alternative. Public purchase of the flood plain would have been another way of spending the $20 million that went to Lucky Peak, but the public benefits would have been intangible and difficult to calculate.

Boise State University Library
**Figure 8.2**
**Most of the Greenbelt underpasses were damaged during the 1983 flood.**

Another incident illustrates the marginalization of value often given to public goods. During the 1983 flood, the Civil Defense coordinator commented that the Greenbelt had "caught hell," but that was what the Greenbelt was designed to do, to "soak up some of this destruction rather than have it on personal property."[15] In fact, that was not at all what the Greenbelt had been designed to do, but the comment indicates a relative depreciation of the investment made by the public for public use compared to investments made by private developers and homeowners. This example and the progress of flood control generally on the Boise River tends to illustrate the paramount role that economic development has played and continues to play in the evolution of a good as fundamental to the public interest as flood safety.

**Figure 8.3**  *Susan M. Stacy*
**After the 1983 flood, the Boise Park Department sponsored revetments to stabilize edge of bike path along the Greenbelt.**

## Western Water Use

Finally, the claims on the water of the Boise River—as on other western waters—have changed in the last 40 years. As mundane a document as the *Boise River Regulation Manual* marks the evolution of priorities in this arid climate. In 1947 Everett Rising, irrigation lobbyist for the Southwest Idaho Water Conservation Project (SWIWCP), gave a speech ranking the appropriate uses of water (after domestic and stock use): first for irrigation, then for industry and navigation. He lumped fish, wildlife, and recreation together as side issues that "should be preserved and protected in proper balance in the accomplishment of [the preferred] uses."[16] At the time he and many other irrigation advocates worried that hydropower projects were going to interfere with this ranking. He lobbied for federal laws to prevent this by placing hydropower close to the bottom of the list. His list did not even mention flood control, an omission in keeping with the general view among SWIWCP members that the Corps of Engineers projects ought to serve reclamation.

In any case, Rising might as well have been ranking the political strength of local constituencies and the pecking order among administrative agencies that followed from it. The priority list in the 1985 revision of the *Boise River Regulation Manual* is a longer list now, definitely including flood control, and also stream flows for fish, and municipal and industrial interests. Federal agencies that did not exist in 1947 have interests in the river and reflect the public attitudes that helped create the Environmental Protection Agency, strengthen the Fish and Game Department, and support agencies such as the Boise Public Works Department. Irrigators have had to confront the reality of these other interests as well as their own needs for funds to maintain and repair their irrigation delivery systems. They have learned to work with the Federal Energy Regulatory Commission in addition to their traditional standard-bearer, the Bureau of Reclamation, an agency with a gradually diminishing mission because new irrigation projects have dwindled.

Federal flood-control policy as implemented on the Boise River has always supported economic development and has always favored structural solutions to flood control. Writer Rochelle Stanfield observed in 1986 that the Corps of Engineers' commitment to nonstructural flood controls had brought the agency to a "new era" in flood control.[17] But where the Boise River rises, Boise citizens, Boise leaders, and Boise agencies will have to make the same commitment before that new era truly dawns.

*129*

*Idaho State Historical Society #82-107.3*

**Figure 8.4**
**The Lucky Peak "rooster tail" in 1958.**

# Appendices

## Appendix A

The report reprinted below, "Boise River: High-Water Years of the Past," was published as Reference Series Number 879 by Merle Wells of the Idaho State Historical Society. Wells based the report on work compiled and prepared by William E. Welsh on June 26, 1944.

### Boise River: High Water Years of the Past

The following statement contains information indicating that before 1895, when the first records of flow of the Boise River were kept, there were several exceedingly high-water years in which floods of catastrophic proportions occurred. This information is based upon precipitation records for the town of Boise, which dates back to 1864, and on statements and stories told by early settlers and old-timers living along the Boise River who had witnessed many of these floods.

Attached to the statement are tabulations showing the precipitation records by month for the town of Boise for the years discussed, as well as 1896 and 1943, which were two of the highest-water years of record; and a table showing the date and quantity (in cubic second feet) of maximum runoff in the Boise River between 1895 and 1943.

It is known that the relationship between precipitation at Boise and precipitation over the Boise watershed is not a constant one; however, Boise is the only station from which records are available dating back to the period for which these records are desired. A notable example of the discrepancy is 1943, when the precipitation over the Boise River watershed was relatively much heavier than that for the city of Boise.

It will be noted that there are a few years of unusually heavy precipitation that were not reported by the early settlers as high-water years on the Boise River: for example 1872 and 1874. However, every year of heavy precipitation is not necessarily a flood year, the greatest single influence on flooding being the weather conditions during the spring months.

Precipitation records for the city of Boise, when considered with statements made by early settlers along the Boise River and information from other sources, indicate that on several occasions during the early years of settlement along the river, the flow of the river was much greater than at any time during the past forty-nine years of record [to 1944]. It will be seen from the tabulation of river flow that the highest flow of record occurred on June 14, 1896, when the river reached 35,500 second feet. Available information from various sources seems to indicate that there were several years, at least one in each decade prior to 1896, in which the flow exceeded that of June 14, 1896.

**1894.** It is my understanding that 1894 was a high-water year throughout the northwest area. Old-timers on the Payette River and the upper Snake River have so indicated. However, I have talked to only two men who have mentioned 1894 as a high-water year on the Boise. Mark Carlyle, who lives across the river from and about one mile west of Parma, insists every time I talk with him that the flow in 1894 was considerably higher than that of 1896. John McGrath, former sheriff of Ada County, moved into his home in the fall of 1894. He said that the high-water marks of that spring were still plain and easily followed. They indicated, he said, that 1894 was a year of considerably higher flow than 1896: that the water had stood around his house and barn. Precipitation records for the town of Boise do not indicate that 1894 was one of our extremely high years, but it is possible that weather—especially high temperatures—during the spring months could have caused a quick, flashy runoff.

**1884.** Precipitation records for Boise indicate that 1884 was an unusually high-water year. This is confirmed by stories from early settlers. A.J. Wiley, now deceased, who formerly resided in Boise and who enjoyed a world-wide reputation as a consulting engineer, made the statement to W.H. Tuller, project manager of the Boise Project, that 1884 was the highest year he had known on the Boise River. During the spring of 1884 Wiley was working on the river where the present Barber Dam is located above Boise.

William Siebenberg, now deceased, who lived on the right bank of the Boise River about three miles west of Caldwell, also stated that 1884 was the highest year that he remembered, higher considerably than any of the years in the '90s. He remembered 1884 particularly because the railroad was built through Caldwell that year.

Jeff Shelton, now deceased, who lived at the east edge of the village of Star, also told W.G. Steward, former assistant engineer of the U.S. Bureau of Reclamation and instructor at the University of Idaho, and the writer, that he especially remembered 1884 as being a year of unusually high water.

**1871.** The precipitation records for the town of Boise indicate that 1871 was the year of highest precipitation since records were started in 1864. W.G. Steward, in his report entitled "Possibility of Flood Damage Along the Boise River Between Diversion Dam and Snake River," says:

> The rainfall records taken at Boise in 1871 would indicate that the flood of that year could easily have been twice the 1896 discharge and possibly three times as great. Computations made from the 1871 precipitation records indicate that the total runoff for that year would have been about 5,500,000 acre feet.

Jeff Shelton also said that there was a year in the early 1870s when he traveled by rowboat from the foot of the hill at the Star cemetery to the porch of the Swalley house, one-half mile above Star. Swalley stated that the only land not under water in that section was the ridge on which his house stood. John Lankford, who lives about one-and-one-half miles west of Star near the bank of the river, said that in 1871 the river broke through near the Canyon County Water Company headgate and covered practically all of the bottom lands from there to Middleton. Former Governor C. Ben Ross, who was born and raised on the farm where he now resides across the river from the village of Parma, says that his father always talked about 1871 as being the highest high-water year.

**1862.** While precipitation records for the town of Boise date back only to 1864, indications from old-times' stories are that the flow in 1862 probably exceeded the high water of any year since that time. I.N. Coston, who . . . homesteaded on a ranch located near the present site of Barber, came to the Boise Valley in 1862. He stated that all land in the river bottoms extending from bluff to bluff and from the present site of Boise westward to the canyon near the present site of Caldwell was completely under water on the 4th of July that year. Jeff Shelton accompanied W.G. Steward and the writer down to Canada Lane, a mile west of Star and then north more than one-half mile from Highway 44 to show us a ridge where a log was found by the first settlers of the Boise Valley (who came in 1864); all were of the opinion that the only way this log could have gotten there was to have floated in by high water from the Boise River.

Fred McConnell, now deceased, a civil engineer and graduate of the University of Idaho, who was born on McConnell Island near the mouth of the river and who spent his life in engineering practice in Canyon County, was a profound student of the Boise River and the various problems involved. He was firmly convinced that there was extremely high water many times during the early days of settlement of the valley; and as further indication of the high water in 1862, he stated to me that his father was with the first emigrant train to travel from the Boise Valley to the Payette Valley, making the descent into Payette Valley on the ridge to the east of Freezeout Hill, as indicated by the monument by the side of the highway. Before starting on the trek across the desert between the two valleys, the party camped north of but near the present town of Eagle, close to Dry Creek and near the bluff, for about three weeks. All the time they were camped there, they burned driftwood which they all believed to have come from high water from the Boise River. This was in the fall of 1862; the party while camped there was visited by the Grimes party, who were returning from the Boise Basin after the discovery of gold on what is now Grimes Creek. An interesting sidelight to this story is the fact that the Grimes party debated the advisability of telling the McConnell party about the discovery of gold in the Boise Basin and decided not to tell it, because there were so many in the McConnell party. However, like all stories of gold discovery in those early days, someone did tell it, and news spread with the resultant gold rush to the basin in the following spring of 1863.

In the National Research Council, American Geophysical Union Transactions of 1941, prepared by the National Research Council of the National Academy of Sciences, Washington, D.C., 1941, there is an article entitled "A Hundred-Year Record of Truckee River Runoff Estimated from Changes in Levels and Volumes of Pyramid and Winnemucca Lakes," by George Hardman and Cruz Venstrom. On page 74 is found this statement: "One of the greatest floods in the history of Nevada occurred in 1861-62." On pages 85 and 86 is a table showing for the period 1839-40 to 1929-30, among other things, a Truckee River Basin precipitation index which shows an index for 1861-62 of 215, considerably the highest shown for any year. On page 88 is a graph showing the estimated runoff of Truckee River, also with 1862 as the highest year of record.

I also have a tracing prepared by Lynn Crandall, Idaho Falls, showing the annual flow into Great Salt Lake. Although this does not show 1862 as the highest year of record, it shows it as one of the highest, exceeded only by 1868, 1864, and 1907. The information from these two sources does seem to indicate that 1862 was a year of unusually high runoff through the entire intermountain area, thus confirming the statements of the early settlers along the Boise River. In fact, there is little doubt in my mind that the flood of 1862 was at least four times the amount of the flood of 1943 and probably much greater [100,000 second feet or greater.]

**1868.** In addition to the foregoing statements from early settlers along the river relative to high-water years, it was thought that it might also be interesting to students of Boise River runoff to know that there was a year of extreme shortage during the late '60s. Mr. and Mrs. Swalley moved to their homestead as bride and groom in the early '60s, either 1864 or 1865, and continued to live there until their deaths a few years ago. Swalley stated to W.G. Steward that there was a year in the late '60s in which the river flow was so low during the summer months that it was possible for a person to wade across the rocks and not get his ankles wet. Swalley stated that at that time there was only one ditch of any consequence taking water out of the river above his farm. Swalley could not remember the exact year but the precipitation for the city of Boise was only 3.52 inches for the year 1868—which was less than one-half the precipitation in Boise for the short-water year of 1924, which was our shortest year of record.

Anton Deiderichsen, who lives about two miles above Payette on the north side of the river on a slough about a mile from the main channel of the river, stated to F.A. Tolman, field engineer, State Department of Reclamation, on August 26, 1947, that he had been on the place since 1891, and that in 1894 they had normal high water in March but that the real high river stage lasted from June into July, and that the preceding snow in Long Valley was fifteen feet deep on the level. That runoff cut the river channel four feet deeper.

Precipitation records for the city of Boise for a few of the extremely high water years, and for the period from October 1 to September 30 of each following year:

|  | 1871/72 | 1884/85 | 1894/95 | 1896/97 | 1943/44 |
|---|---|---|---|---|---|
| October | .24 | 4.06 | .84 | .00 | .58 |
| November | 2.13 | .46 | 3.14 | .71 | 2.31 |
| December | 1.26 | 2.27 | .48 | .69 | 2.75 |
| January | 3.54 | 1.75 | 2.88 | 3.72 | 1.27 |
| February | 1.29 | 1.32 | .82 | .50 | .88 |
| March | 7.66 | 2.78 | 1.64 | 2.41 | 2.08 |
| April | 1.54 | .78 | 1.26 | 2.72 | 1.39 |
| May | 2.75 | .92 | 2.08 | 4.90 | .72 |
| June | .64 | 3.41 | .19 | 1.36 | 1.09 |
| July | .14 | .60 | .00 | .10 | .16 |
| August | .00 | .07 | .06 | 1.06 | — |
| September | .11 | 2.11 | .40 | .28 | — |
|  | 21.30 | 20.53 | 13.79 | 18.45 | 13.23 |

## Boise River Runoff, 1895-1943

| Year | Maximum Discharge | Date of Maximum | Year | Maximum Discharge | Date of Maximum |
|---|---|---|---|---|---|
| 1895 | 7,880 | 5/6 | 1920 | 9,623 | 5/18 |
| 1896 | 35,500 | 6/14 | 1921 | 19,682 | 5/18 |
| 1897 | 29,500 | 4/19 | 1922 | 18,174 | 5/26 |
| 1898 | 6,540 | 4/26-28 | 1923 | 11,950 | 5/26 |
| 1899 | 19,000 | 5/15 | 1924 | 5,186 | 5/18 |
| 1900 | 11,960 | 5/12 | 1925 | 14,350 | 5/20 |
| 1901 | 12,700 | 5/16 | 1926 | 7,094 | 6/6 |
| 1902 | 8,190 | 5/29 | 1927 | 20,061 | 5/18 |
| 1903 | 16,800 | 6/2 | 1928 | 20,710 | 5/10 |
| 1904 | 19,700 | 4/15 | 1929 | 7,518 | 5/30 |
| 1905 | 6,260 | 6/2 | 1930 | 7,599 | 5/30 |
| 1906 | 8,710 | 5/12 | 1931 | 5,434 | 5/8 |
| 1907 | 17,000 | 4/15 | 1932 | 13,580 | 5/14 |
| 1908 | 10,600 | 5/22 | 1933 | 12,214 | 6/16 |
| 1909 | 16,000 | 6/5 | 1934 | 5,274 | 4/14 |
| 1910 | 16,600 | 3/2 | 1935 | 9,501 | 5/25 |
| 1911 | 15,100 | 6/15 | 1936 | 19,790 | 4/24 |
| 1912 | 15,600 | 6/9 | 1937 | 7,705 | 5/6 |
| 1913 | 13,300 | 5/28 | 1938 | 19,283 | 5/2 |
| 1914 | 11,300 | 4/16 | 1939 | 8,413 | 5/1 |
| 1915 | 6,227 | 5/19 | 1940 | 9,866 | 5/14 |
| 1916 | 16,550 | 5/7 | 1941 | 8,861 | 5/27 |
| 1917 | 17,848 | 5/15 | 1942 | 10,690 | 5/27 |
| 1918 | 12,601 | 6/14 | 1943 | 25,040 | 4/18 |
| 1919 | 11,580 | 5/30 | | | |

## Appendix B

## Annual Maximum Mean Daily Discharge of Boise River*
(cfs)

| Year | Unregulated at Diversion Dam | Regulated at Diversion Dam | Regulated at Boise |
|---|---|---|---|
| 1944 | 7,632 | | |
| 1945 | 11,644 | | |
| 1946 | 18,812 | | |
| 1947 | 13,838 | | |
| 1948 | 15,260 | | |
| 1949 | 12,829 | | |
| 1950 | 13,673 | | |
| 1951 | 14,065 | | |
| 1952 | 23,429 | | |
| 1953 | 14,790 | | |
| 1954 | — | | |
| 1955 | 10,480 | 5,110 | 1,740 |
| 1956 | 22,950 | 9,470 | 6,840 |
| 1957 | 16,930 | 10,600 | 6,870 |
| 1958 | 21,750 | 10,000 | 6,320 |
| 1959 | 9,040 | 5,390 | 1,800 |
| 1960 | 11,840 | 8,200 | 5,710 |
| 1961 | 7,830 | 5,360 | 1,560 |
| 1962 | 11,340 | 5,320 | 1,540 |
| 1963 | 11,480 | 9,820 | 5,870 |
| 1964 | 10,940 | 7,230 | 4,630 |
| 1965 | 20,850 | 11,600 | 7,170 |
| 1966 | 8,220 | 4,960 | 1,760 |
| 1967 | 15,600 | 5,270 | 1,640 |
| 1968 | 7,050 | 5,130 | 1,800 |
| 1969 | 15,930 | 8,660 | 5,280 |
| 1970 | 14,850 | 8,500 | 5,030 |
| 1971 | 20,250 | 10,800 | 6,850 |
| 1972 | 19,600 | 10,200 | 6,710 |
| 1973 | 9,550 | 4,760 | 1,460 |
| 1974 | 18,500 | 10,815 | 7,350 |
| 1975 | 20,618 | 10,800 | 8,130 |
| 1976 | 13,732 | 10,500 | 6,900 |

*Idaho Department of Water Resources, *Review of Boise River Flood Control Management* (Boise, 1974), p. 17; also, staff at Department of Water Resources. "At Diversion Dam" refers to flow into Diversion Dam pool. "Unregulated flow" refers to Lucky Peak outflow corrected for storage changes at upstream reservoirs.

| Year | Unregulated at Diversion Dam | Regulated at Diversion Dam | Regulated at Boise |
|------|------------------------------|----------------------------|--------------------|
| 1977 | 3,190 | 3,870 | 1,290 |
| 1978 | 11,896 | 7,240 | 5,460 |
| 1979 | 9,620 | 4,820 | 1,560 |
| 1980 | 12,960 | 9,180 | 5,650 |
| 1981 | 9,710 | 9,500 | 6,880 |
| 1982 | 18,320 | 10,700 | 7,500 |
| 1983 | 24,290 | 13,200 | 9,840 |
| 1984 | 22,430 | 10,500 | 6,950 |
| 1985 | 9,120 | 4,750 | 2,420 |

# Notes

### Abbreviations

*IDS*: Refers to the *Idaho Statesman*. This Boise-based newspaper was named *Idaho Tri-Weekly Statesman* from 1864 to 1884, *Idaho Daily Statesman* from 1884 to 1969, and the *Idaho Statesman* after 1969.

IHS: Idaho Historical Society, Boise, Idaho. The Society's library is the depository for state documents and other manuscript collections.

M-K RC: Morrison-Knudsen Records Center, Boise, Idaho.

SRC: Federal Records Center in Seattle, Washington.

### Introduction

1. "Formal Dedication of Lucky Peak Dam June 23 To Include Demonstration of Equipment at Work," *IDS*, June 3, 1955, p. 18.

2. Martin Reuss, "Andrew A. Humphreys and the Development of Hydraulic Engineering: Politics and Technology in the Army Corps of Engineers, 1850–1950," *Technology and Culture* 26 (January 1985), p. 30.

3. Keith W. Muckleston, "The Evolution of Approaches to Flood Damage Reduction." *Journal of Soil and Water Conservation* 31 (March–April 1976), p. 53. Muckleston quotes J.E. Goddard, "An Evaluation of Urban Flood Plains," American Society of Civil Engineers, New York, 1973, and G.F. White and J.E. Haas, *Assessment of Research on Natural Hazards* (Cambridge: MIT Press, 1975). See also Jamie W. Moore and Dorothy P. Moore, *The Army Corps of Engineers and the Evolution of Federal Flood Plain Management Policy* (University of Colorado:

Natural Hazards Research and Applications Information Center, 1989), p. 122–28.

4. House Committee on Public Works, *A Unified National Program for Managing Flood Losses, Communication from the President of the United States Transmitting a Report by the Task Force on Federal Flood Control Policy*, H. Doc. 465, 89th Cong., 2nd sess., 1966.

5. For a discussion of nonstructural alternatives, see *Seminar Proceedings, Implementation of Nonstructural Measures*, Policy Study 83–G520 (Washington, D.C.: Army Corps of Engineers, 1983).

6. Public Law 93–251, Section 73.

7. Jeffrey Stine, "Environmental Politics and Water Resource Development: The Case of the Army Corps of Engineers During the 1970s" (Ph.D. dissertation, University of California, Santa Barbara, 1984), p. 187.

8. "Flood Experts Call for Curbs on Development in Low-Lying Areas," B. Drummond Ayres, Jr., *New York Times*, May 27, 1973, p. 40. For typical examples of articles on nonstructural projects, see: Marvin Zeldin, "Corps New Look in Flood Control: No Dams, Levees," *Audubon* 77 (July 1975), p. 103–104; Brent Blackwelder, "In Lieu of Dams," *Water Spectrum* 9 (Fall 1977), p. 41–46; Frank Notardonato, "Corps Takes New Approach to Flood Control," *Civil Engineering* 49 (June 1979), p. 65–68; James Nathan Miller, "A Plan to Use our Floods, Not Fight Them," *Readers Digest* 102 (March 1973), p. 21–22, 24–26; John Chandler and Arthur Doyle, "An Alliance With Nature," *Water Spectrum* 10 (Summer 1978), p. 24–31.

9. Donald Worster, "New West, True West: Interpreting The Region's History," *The Western Historical Quarterly* 18 (April 1987), p. 152.

10. Scott W. Reed, "The Other Uses for Water," *Idaho Yesterdays* 30 (Spring/Summer, 1986), p. 44. Reed discusses the evolution of water law in Idaho and the likely impact of the California Supreme Court Mono Lake decision on the "public trust" doctrine. Reed concluded, "Mother Nature establish[ed] multiple uses for water millennia ago; this diversity, finally recognized by the legislature and the courts will make it extremely difficult to continue to claim 99% for the single purpose of irrigation."

11. Charles McKinley made this observation in *Uncle Sam and the Pacific Northwest* (Berkeley: University of California Press, 1952), p. 110–111. He described the sharp conflict between the Corps and the Fish and Wildlife Service over the pace of dam construction on the mid-Columbia in the late 1940s. The two agencies tried to cooperate with one another, but the interests "expressed through" the Corps, Bonneville Power Administration, and the Bureau of Reclamation were too powerful to "stay the developments that menace the salmon runs and the commercial and sports fisheries dependent thereon, for whom the Fish and Wildlife speaks."

12. A vast political science literature discusses the role of federal agencies in the policymaking process and their relationships with national interest groups. Two general sources are John W. Kingdon, *Agendas, Alternatives, and Public Policies* (Boston: Little, Brown, and Co., 1984) and Francis E. Rourke, *Bureaucracy, Politics, and Public Policy* (Boston: Little, Brown, and Co., 1984.) Rourke includes an extensive bibliography. See also: Jeffrey K. Stine and Michael C. Robinson, *The US Army Corps of Engineers and Environmental Issues in the Twentieth Century* (Washington: Historical Division, Office of Administrative Services, Office of the Chief of Engineers, 1984).

13. Quoted frequently, the remark appeared in a column by George F. Will published in the *Idaho Statesman*, June 2, 1988, p. 10–A.

## Chapter One

1. "Aimed at Flood Control, Costly Lucky Peak Dam Could be Biggest Hoax; When Were the Floods?" *IDS*, August 2, 1953, p. 4. See also "We Are Told We Have the Wrong Information, So We Correct an Opinion of Arrowrock Dam," *IDS*, August 16, 1953, p. 4.

2. U.S. Army Corps of Engineers, *Review of Survey Report, Boise River, Idaho, With a View to Control of Floods* (Portland: District Office, 2 January 1946), p. 16. Occasionally, floods are caused by unusual storms, discussed at the same page. (Hereafter referred to as *Review Survey Report*.)

3. Eugene Chaffee, "Boise, the Founding of a City," *Idaho Yesterdays* 7 (Summer, 1963), p. 3. See also Chaffee, "Early History of the Boise Region, 1811–1864," (Masters thesis, University of California, 1971), p. 62–63.

4. Merle Wells, *Boise, An Illustrated History* (Woodland Hills, California: Windsor Publications, Inc., 1982), p. 19. Mr. Davis later gave some of his land to Boise in memory of his wife Julia, now Julia Davis Park.

5. The third water right, for example, was issued to the Middleton Irrigation system, and carried water to farms on 4,400 acres. Neil H. Carlton, "A History of the Development of the Boise Irrigation Project," (Masters thesis, Brigham Young University, 1969), p. 11.

6. *IDS*, May 4, 1876, p. 2.

7. *IDS*, May 2, 1876, p. 3. The shifting of the channel was perceived as an opportunity by George Washington Stilts, a prominent blacksmith of the town, to stake claims in the old channel, hoping to find gold. Kelly wrote, "We can see no case for the failure of these enterprises." *IDS*, June 6, 1876, p. 3. For stories about Stilts, see Thomas Donaldson, *Idaho of Yesterday*, (Caldwell, Caxton Printers, 1941; reprinted by Greenwood Press of Westport, Conn., 1970), p. 113–120.

8. *IDS*, May 4, 1876, p. 2.

9. *IDS*, May 4, 1876, p. 2.

10. *IDS*, May 9, 1876, p. 2.
11. *IDS*, May 11, 1876, p. 2.
12. *IDS*, May 16, 1876, p. 3, and May 25, 1876, p. 3.
13. *IDS*, December 16, 1876, p. 3. The bridge was replaced in 1912. Now known as the Eighth Street Bridge, the Ada County Highway District closed it to vehicular traffic in 1987; it is now used only by pedestrians and bicyclists.
14. *IDS*, May 16, 1876, p. 3.
15. See Appendix A for a table of flood discharge estimates by the Corps of Engineers.
16. *Review Survey Report*, Table 2.
17. Wells, *Boise, An Illustrated History*, p. 40.
18. *IDS*, June 3, 1896, p. 4.
19. *IDS*, June 2, 1896, p. 3.
20. *Caldwell Tribune*, June 6, 1896, p. 2.
21. *IDS*, June 13, 1896, p. 6.
22. Helen Lowell and Lucile Peterson, *Our First Hundred Years, A Biography of Lower Boise Valley, 1814–1914* (Caldwell, Idaho: Caxton Printers, 1946), p. 94.
23. Carlton, p. 89.
24. Carlton, p. 124–125.
25. Carlton, p. 103.
26. *Review Survey Report*, p. 42.
27. Board of Engineers for Rivers and Harbors, *Survey Report of Boise River*, 76th Cong., 3rd sess., H. Doc. 957, September 23, 1940. The report considered possible improvements to the channel, levees, and dikes, but concluded that complete protection was not economically justified. The most suitable improvement plan was to build Anderson Ranch Reservoir.
28. *Review Survey Report*, p. 20. Also see United States Bureau of the Census, *Fourteenth Census of the United States*, and Howard R. Lamar, *Reader's Encyclopedia of the American West* (New York: Thomas Y. Crowell, 1977), p. 1000–1003, for a brief overview of the reclamation movement in the West.
29. A Corps of Engineers Aerial Photograph dated May 1950 (see cover of this book) shows the river channel around 1909 and the improved channel to the south installed by Boise City. The date of the straightening was not included on the map, but was probably done in the 1920s or 1930s. Some of the earliest fill work was at the site of the present Boise State University campus. The river had two main channels with islands and connecting channels between them. The work prepared the site for the city's first municipal airport in 1928 by filling the south channel and leveling the islands for the runway. See Eugene B. Chaffee, *Boise College, An Idea Grows* (Boise, 1970), p. 55–56.
30. Before 1936 Congress gave the Corps flood-control responsibilities in the Mississippi and Sacramento River Valleys. See Flood Control Act of 1917, P.L.

367, 64th Congress, 1 March 1917 (39 Stat. 949, U.S.C. 703), and Flood Control and Protection Act of 1928, P.L. 391, 70th Congress, 15 May 1928 (45 Stat. 534, 33 U.S.C. 702a).

31. Idaho ranked 4th in public works grants from the New Deal, having suffered protracted agricultural depression since the 1920s, emigration, loss of income, decline in demand for timber and mining products, and extreme drought. See Leonard J. Arrington, "Idaho and the Great Depression," *Idaho Yesterdays* 13 (Summer, 1969), p. 2–8.

32. U.S. Department of the Army, *Appendix B, Hearing, Review of Survey Report, Boise River, Idaho, With A View to Control of Floods. Proceedings of Hearing 14 Dec 1944* (Portland: District Office, 2 January 1946), p. 49–50. (Hereafter referred to as *Appendix B*.) Testimony at this 1944 public hearing detailed the productivity and value of the river bottom lands now put to row crops such as beets, potatoes, beans, onions, and others. Canyon County, given an "A" award for food production for the war effort, had produced 25% of the country's seed for lima beans in 1942, 38% of onions, and 60% of hybrid sweet corn.

33. This figure was derived by subtracting the discharge for Arrowrock Reservoir from the natural discharge for each year and then averaging. The data is in Table 2 in the *Review Survey Report*. The mitigating effect of Arrowrock was most significant in the year 1917 when the dam took the edge off the flood by 8,150 cfs. Two years later the effect was only 780 cfs, the lowest difference of the 29-year period. See also p. 15 of *Review Survey Report*.

34. *IDS*, April 3, 1946, p. 1. A watermaster is an individual responsible by law to supervise the distribution of a stream's water according to established rights.

35. *Appendix B*, 1946, p. 164–168. This testimony was given by Paul Davis, the Secretary of the Ada County Flood Control Committee. He described in detail the organization efforts, and concluded with pride, "In being prepared, the old American individualism and ingenuity prevailed and our Boise Valley people took their misfortune in stride."

36. U.S. Corps of Engineers, *Downstream Channel Requirements, Lucky Peak Dam and Reservoir, Boise River, Idaho*, Design Memorandum #6 (Walla Walla: District Office, 1958), p. 6. (Hereafter referred to as *Downstream Channel Requirements*.)

37. *Review Survey Report*, 1946, Appendix A, Table 9. Accounts vary as to the amount which passed through Boise. William Welsh said in a *Statesman* interview on May 28, 1943, that the discharge on April 17 was 23,105 cfs; he may have referred to the discharge above the New York Canal diversion.

38. Both domestic and wild animals crowded together on one high piece of ground on Eagle Island downstream from Boise. When the farmer realized his opportunity, he brought dogs onto the narrow strip, where they proceeded to kill a thousand gophers. *IDS*, June 15, 1983, p. 1-A.

39. *Caldwell News Tribune*, April 22, 23, 24, 1943. Results of any subsequent investigations were not reported.

40. To Adjutant General, State of Idaho, from Lt. Col. Ralph A. Breshears, May 3, 1943, Papers of Governor C.A. Bottolfsen, IHS, Box 1, "Flood Control."

41. In earlier years farmers moved cattle to higher ground and let them browse the sagebrush until the flood receded. In 1943 dairy herds occupied the flood plain, and these needed pasturage and hay. Transcript of meeting at Governor's office on March 26, 1943, Papers of Governor Bottolfsen, IHS, Box 1, "Flood Control."

42. *Review Survey Report*, p. 29.

43. *Appendix B*, p. 29.

44. *IDS*, April 16, 1943, p. 1. Boise City sewer lines discharged directly into the river. The city's first sewage treatment plant was built at Lander Street in 1949–50.

45. Oliver Lewis, Corps Area Engineer, to Gov. C.A. Bottolfsen, April 30, 1943, Papers of Governor Bottolfsen, IHS, Box 1, "Flood Control."

46. Col. Ralph Tudor, Portland District Engineer, to Senator Henry Dworshak, May 28, 1943, Papers of Governor Bottolfsen, IHS, Box 1, "Flood Control." Although a report was prepared and sent to Dworshak, a copy of it was not in either Dworshak's or Bottolfsen's papers or among the other Corps of Engineers' papers consulted. See letter to William Tuller from "R," dated August 23, 1943, Papers of Henry Dworshak, IHS, Box 14, "Interior–Reclamation–Anderson Ranch–1943."

47. *Appendix B*, p.1–2. The Corps did not detail its methodology in performing the survey. Minor discrepancies exist between it and the testimony of witnesses.

48. *Review Survey Report*, p. 31. The Corps estimates are further broken down into agricultural, irrigation and drainage structures, and siltation. I have combined these items as "agricultural."

## Chapter Two

1. "Talk Widening Boise River, Consider Straightening Channel To Avoid Floods," *Caldwell News Tribune*, April 29, 1943, p. 1.

2. William Welsh to Col. Donald J. Leehey, Papers of Henry Dworshak, IHS, Box 14, "War Dept–Flood Control–1943."

3. "House Refuses to Allow Fund for Boise River," *IDS*, January 27, 1937, p. 5.

4. *Appendix B*, p. 14–16.

5. See *Appendix B*, p. 119. Even in the testimony at this hearing the debate about the causes of turbid water went on, with one speaker blaming siltation on timber practices; another, mining practices. Still another disputed the fact that turbid water was bad for fish and claimed that the problem with murky water was that it was just harder to catch them. In 1950 the Department of Agriculture produced a report on the conditions of the Boise Front entitled *Survey Report, Boise River Watershed*. The report cited specific practices in mining, timbering, irrigation diversion,

grazing, and other enterprises that each contributed one way or another to the exacerbation of flooding problems.

6. *Appendix B*, p. 135.

7. Although there was much interest in levees, it is possible that the fill and levee work already done in Boise may have been partly responsible for some of the aggravated flooding that occurred downstream. In a map showing the boundaries of the floods of 1896, 1936, and 1943, *Review Survey Report*, Plate IV, the floods are shown contained close to the river banks at several places in the urban area especially during the later floods. This may have contributed to an accelerated velocity which exacerbated bank erosion downstream.

8. *Appendix B*, p. 61–62.

9. Boise City, "A Policy Plan for the Boise Metropolitan Area," Boise River Plan Amendment, Ordinance 4863; adopted September 30, 1985. See Recreation Policy 1e.

10. William Welsh to Governor Bottolfsen, April 22, 1943, Papers of Governor Bottolfsen, IHS, "Department of Reclamation, 1943."

11. To Congressman Henry Dworshak from Major General Thomas M. Robins, Acting Chief of Engineers, 18 August 1943, Papers of Henry Dworshak, IHS, Box 14, "Interior–Reclamation–Anderson Ranch–1943."

12. United States, Congress, House, H. Doc. No. 308, 69th Cong., 1st sess.; enacted into law as Section 1 of the River and Harbor Act on January 21, 1927.

13. See United States, Congress, House, *Report of the US Army Corps of Engineers on the Columbia River and Its Minor Tributaries*, House Document 103, 73rd. Congress, 1st. sess., 1933, p. 3; United States, Congress, House, *Report of the US Army Corps of Engineers on the Snake River and its Tributaries*, House Document 190, 73rd. Congress, 2nd. sess., 1934, p. 6. (Hereafter referred to as H. Doc. 103 and 190 respectively.) See also Jamie W. Moore and Dorothy P. Moore, *The Army Corps of Engineers and the Evolution of Federal Flood Plain Management Policy* (University of Colorado: Natural Hazards Research and Applications Information Center, 1989); and Beatrice H. Holmes, "Federal Participation in Land Use Decision-Making at the Water's Edge—Floodplains and Wetlands," *Natural Resources Lawyer* 13 (No. 2, 1980), p. 351–410.

14. Flood Control Act of 1936, 49 Stat. 1570.

15. McKinley's *Uncle Sam and the Pacific Northwest* provides an excellent discussion of the role of the Corps and the Bureau in water resource planning during this period. See Postscript, p. 636–661. See also Chapters II, III, and XVIII.

16. James B. Weatherby, "The Hells Canyon Controversy: A Study of the Hells Canyon Associations and their View of Comprehensive River Basin Development" (Masters thesis, University of Idaho, 1968), p. 11–12.

17. See United States, Army Corps of Engineers, *The History of the North Pacific Division, US Army Corps of Engineers, 1888 to 1965* by Roy W. Scheufele

(Portland: North Pacific Division, n.d.), p. 13, 24. (Hereafter referred to as *North Pacific Division*.) The harmonized Bureau of Reclamation report was printed as H. Doc. 473, referred to in note # 7 in this chapter.

**Chapter Three**

1. Copies of the telegrams dated October 18, 1943, from H. W. Morrison to Senators John Thomas and D. Worth Clark and Representatives Henry C. Dworshak and Compton I. White; from Governor C.A. Bottolfsen to Congressional delegation; and from D. Worth Clark to Governor C.A. Bottolfsen, October 19, 1943, are all in Papers of Governor Bottolfsen, IHS, Box 1, "Flood Control." In addition to the study on Lucky Peak, Congress authorized $460,000 in assistance for reconstruction of levees damaged by the flood.

2. Governor Robert Smylie, interview with author, in Boise, April 29, 1991.

3. Address by Harry W. Morrison before the annual meeting of the Idaho State Reclamation Association at Twin Falls May 18, 1940, entitled "Southwest Idaho Water Conservation Project and Its Objectives," Papers of Henry Dworshak, IHS, Box 14, "Interior–Reclamation–Misc–1944."

4. J.L. Driscoll to Senator Herman Welker, November 15, 1950, M-K RC, Box S6: 5–41, File "1951–Southwest Idaho Water Conservation Project–General Correspondence."

5. *The Valleys of Tomorrow*, published by SWIWCP in 1940, p. 3. Copy in Papers of Governor Bottolfsen, IHS, "Reclamation, 1939–40."

6. "Future Irrigation Program—Southwestern Idaho," March 1940, Papers of Governor Bottolfsen, IHS, "Reclamation, 1939–40."

7. Glen Barrett, "Reclamation's New Deal for Heavy Construction, M-K in the Great Depression," *Idaho Yesterdays* 22 (Fall 1978), p. 21–27. Numerous letters dated from the 1940s and 1950s, located in papers of Idaho governors and at the M-K RC list officers and executive committee members prominently on SWIWCP stationery.

8. *Valleys of Tomorrow*, p. 5.

9. See McKinley, p. 623. McKinley discusses Rising's connections to private utilities and to the National Reclamation Association. Rising figured in documentation gathered by the Federal Trade Commission in an investigation of the electric utilities' methods of lobbying. See also Glen Barrett, "Reclamation's New Deal for Heavy Construction, M-K in the Great Depression," *Idaho Yesterdays*, 22 (Fall 1978), p. 21–27.

10. United States, Congress, House, *The Columbia River: Supplemental Reports on the Bitterroot Valley, Cambridge Bench, Canby, Council, Crooked River, The Dalles (West Unit), Hells Canyon, Mountain Home (Payette Unit), Upper Star Valley, and Vale (Bully Creek Extension) Projects*, House Document 473, 81st

Cong., 2nd. Sess., Volume 2, p. 221, 1950. (Hereafter referred to as H. Doc. 473, Vol. 2.)

11. United States, Bureau of Reclamation, "Initial Unit, Mountain Home Project, Idaho," Report #69–A.

12. H. Doc. 473, Vol. 2, p. 221–224.

13. H.Doc. 473, Vol. 2, p. 248–249.

14. See weekly newspaper *Statewide*, May 24, 1951, p. 4. While there were other organized advocates of irrigation projects, such as the Idaho Reclamation Association, SWIWCP was perceived as the chief lobby for irrigation. In a May 9, 1951, p. 2 (2nd edition) article in the *Idaho Free Press* on the Hells Canyon controversy, SWIWCP was characterized as "the chief lobbying agency for irrigation development. These men have the most to gain by free enterprise, the most to lose by socialism, but some of them have much to gain by the construction of the dam, irrespective of its ultimate effect on them." Further, in a column in the *Idaho Statesman* on April 29, 1951, p. 7, John Corlett, also writing about Hells Canyon, said SWIWCP had "spearheaded the drive for all irrigation development in southwestern Idaho the last two decades." He added that SWIWCP had believed the Mountain Home project feasible since before World War II. For discussion of irrigation as a perceived growth stimulant, see also Susan M. Stacy, *Legacy of Light, A History of the Idaho Power Company* (Boise: Idaho Power Company, 1991).

15. *Appendix B*, p. 7.

16. The site had been located by Corps engineer Charles Greenwood, who recalled the story in an interview by the *Idaho Statesman* on August 3, 1979. According to Greenwood, he first examined the site early in 1943. (This date was probably reported in error; the site must have been located in 1944.) Geologists had advised Greenwood to look for a dam site on the North Fork of the river because they felt the bedrock further downstream was too porous and soft. But Greenwood was eating his lunch one day just upstream from Diversion Dam at a state park and wondered why the river made a horseshoe bend just above the park. When he looked into the river bed, he saw hard, intrusive lava rock and realized that the river had eroded the gravel lying over the rock and was now going around it. When the District office later confirmed the tightness and solidity of the rock, there was little further discussion of other potential sites.

17. Ron Barrett to Susan Stacy, September 16, 1986. The quad sheet is titled "Lucky Peak, Idaho."

18. *Appendix B*, p. 3.

19. *Appendix B*, p. 198–200.

20. *Appendix B*, p. 3; and "Old Timers Due to Attend Flood Hearings," *IDS*, December 13, 1944, p. 9.

21. E.W. Rising to Harry W. Morrison, October 30, 1945, M-K RC, S6: 5–41.

22. The "unit by unit" strategy adopted by SWIWCP and the Bureau came under criticism in an April 7, 1949, p. 3, article by Ed Emerine, a writer for *Statewide*, a weekly originating in Boise. Reacting to the reported high cost of the tunnel from Garden Valley to the Boise River, he said the Bureau should tell the public just how much it would cost "to water the Mountain Home desert" in total instead of piece-by-piece. He felt that the project should be self-supporting or not funded at all.

23. Major General Thomas M. Robins, Acting Chief of Engineers, to Senator Henry Dworshak, 18 August 1943, Papers of Henry Dworshak, IHS, Box 14, "Interior–Reclamation–Anderson Ranch–1943."

24. *Review of Survey Report, Boise River*. For full citation, see Chapter One, note # 2.

25. *Review Survey Report*, p. 61, 64. The Corps estimated the amount of damage expected from floods of various sizes, determined the frequency of each flood, and added the damages expected over time. The average was $444,810 per year. That was the benefit. The annual cost of the dam was $507,630. Three years after Lucky Peak was built, a Corps report suggested that flood-damage estimates may have been "unreasonably large," meaning that benefits of the dam would have been overestimated also. See *Downstream Channel Requirements*, p. 9–10.

26. This computation is based on 286,000 acre-feet in Arrowrock, 177,000 in Deer Flat, 470,000 in Anderson Ranch, and 280,000 in Lucky Peak.

27. *Review Survey Report*, p. 65. The gross value of supplemental water ranged between $3 and $8 per acre foot in those years it had been needed. The Corps referred to its $1 value as "a nominal net return" per acre foot using the lower figure in this range, the "annual value for supplemental water that might be creditable to a third reservoir in the system."

28. *Review Survey Report*, p. 65–66.

29. *Review Survey Report*, p. 66.

30. *Review Survey Report*, p. 67.

31. Due to the wide and shallow nature of the Boise floods, and because there was usually a warning of several days or weeks, loss of life was rare.

32. *Review Survey Report*, p. 67–69. The dollar values were: benefits, $632,910; and costs, $507,630. The 1:1 ratio was a requirement of the 1936 Flood Control Act.

33. *Review Survey Report*, taken from two tables, p. 70, 71.

34. *Review Survey Report*, p. 71.

35. *Review Survey Report*, p. 72.

36. *Review Survey Report*, p. 82–83.

37. *Review Survey Report*, p. 76.

38. "Big Area Dam May Begin Construction in Summer," *IDS*, May 2, 1946, p. 1.

39. See "Welsh Sees $40,256,000 Flood Damage," *IDS*, June 9, 1946, p. 7, and May 2, 1946, p. 1. In 1949, Welsh accepted a position as Secretary-Manager of the National Reclamation Association, disappointing Harry Morrison, who never felt that "the National Reclamation Association was of much assistance to Idaho, but I appreciate the ramifications of the activity of an association of this sort." To Harry Morrison from William Welsh, February 11, 1949 and to Welsh from Morrison, February 12, 1949, M-K RC, S6: 5–41. The last remarks were made to a meeting of the Columbia River Inter-Agency Committee.

40. "Senate Group Voices Doubt on Local Dam," *IDS*, June 25, 1946, p. 1. See also telephone note, Bousquet, OCE to Col. Walsh, Portland District Engineer, 24 June 1946, SRC, 77–82–0060, Box 1/18, "Boise River Review Report folder #2."

41. Public Law 526, July 24, 1946, 79th Cong., 2nd Sess. (60 Stat. 641, P. 650.) See also *IDS*, June 26, 1946.

42. "Gossett Seeks Funds to Finish Anderson Dam," *IDS*, June 29, 1946, p. 2.

43. United States, Corps of Engineers, *Definite Project Report on Lucky Peak Dam, Boise River, Idaho* (Walla Walla: District Office, October 3, 1949). (Hereafter referred to as *Definite Project Report*.)

44. *Definite Project Report*, p. 78.

45. *Definite Project Report*, p. 80–81. The elimination of power benefits may have been related to a wider discussion underway at the time about power subsidization. In a letter to Lynn Driscoll from B.J. Weis, Harry Morrison's assistant, on April 8, 1947, (M-K RC, S 6: 5–41), Weis wrote that eastern industrialists were beginning to object to such subsidization particularly in the West because it would lead to industrial development and competition with themselves. Funding for reclamation projects thus subsidized was being reduced in 1947 by 53%. Weis said that SWIWCP did not advocate or oppose public power as such, but that it must recognize that western reclamation projects often were perceived as industrial subsidies. The promoters of Lucky Peak, along with the Corps, may have chosen to remove power sale benefits in order not to threaten the project by so doing.

46. See *Definite Project Report*, p. 14–15. Also "Lucky Peak Initial Operations Computations," 26 September 1949, p. 5, SRC, 77–83–0009, Box 3/10. The Corps gave much consideration to the relationships between flood-control operations and irrigation opportunities at various levels of permanent pool. One idea involved reducing active storage to as low as 99,000 acre feet.

47. *North Pacific Division*, p. 25–26. The author did not record the nature of the differences.

48. McKinley, p. 641–42. Copy of the agreement is also in H. Doc. 473.

49. Governor C.A. Robins to Congressman Abe McGregor Goff, Papers of Governor Robins, IHS, Box 8, Department of Reclamation, 1947–1949.

50. E.W. Rising to Harry Morrison, June 27, 1949, M-K RC, S6: 5–41. The *Annual Reports of the Chief of Engineers, U.S. Army Civil Works Activities,* from 1948 through 1955 in chapters on Walla Walla District indicate the following record of expenditures for Lucky Peak:

    1948    $152,867
    1949    $383,417
    1950    $921,826
    1951    $3,178,851
    1952    $2,652,609
    1953    $4,995,512
    1954    $3,670,796
    1955    $1,621,116 and $9,668 for maintenance.

According to the 1970 *Annual Report*, the federal cost of the completed project was $19,081,250, less than the costs projected in the *Definite Project Report*. Another $506,000 was added by the cost of additional recreation facilities. The figures above do not add up to the total because of the continuing recreation and other improvements.

51. Macco-Puget Sound Bridge Dredging constructed the tunnel, Hunt and Willett built the intake structure and outlet diversion, and the Roy L. Bair Co. built the Mores Creek Highway bridge. A list of contractors was printed on the back of the Dedication Program distributed for the Dedication Ceremony on June 23, 1955.

52. Earl Reynolds, interview with author, in Boise, February 17, 1987.

53. E.W. Rising to Harry Morrison and Lynn Driscoll, December 7, 1949, M-K RC, S6: 5–41.

54. *Review Survey Report*, p. A–25–27. The Corps wrote, "After the ultimate development [including the Mountain Home Desert] of the irrigable lands in the Boise watershed, the entire run-off of the Boise River will be utilized for irrigation."

55. B.J. Weis to Lynn Driscoll and William Welsh, April 8, 1947, M-K RC, S6: 5–41. These remarks were contained in a proposed draft statement to be used in the event of eastern industrial resistance to the subsidization of multipurpose projects.

56. E.W. Rising to H.W. Morrison, September 12, 1950, M-K RC, S6: 5–41. Morrison's interests were also affected directly, as well as on behalf of the Mountain Home Project, since his company was in the midst of executing its contract to build the embankment at Lucky Peak.

57. Harry Morrison to A.E. Stoddard, President of Union Pacific Railroad, November 8, 1950, M-K RC, S6: 5–41. This letter explained the accomplishments of SWIWCP to date in bringing about "increased irrigation development," namely, Cascade Reservoir, Anderson Ranch Dam, and "the latest dam we have gotten underway is to be known as Lucky Peak Dam," with which "we will have the Boise

River waters entirely under control and available for beneficial use." He further identified what they still had to accomplish and the rationale for supporting the high dam at Hells Canyon. "We definitely feel that Idaho must maintain its position in the overall irrigation development [of the country] and be prepared to insist on the construction of feasible projects in our state when conditions permit. I sincerely hope that we can depend on your support in a continuing effort to conclude this program which has been so effectively advanced to its present status."

58. "Crowd Attends Dedication for Lucky Peak Project," *IDS*, June 24, 1955, p. 16. Although the *Statesman* did not mention it, Lucky Peak was the first of five dams the Corps dedicated within a span of four months. On June 24, 1955, the *Oregon Journal* referred to the Corps achievement in the Northwest as the "dam-a-day dedication program." The other dams were Albeni Falls at Pend Oreille, Lookout Point on the Willamette, Chief Joseph on the Columbia, and the Dexter on the Willamette. "This is the greatest array of large dams ever to be completed and placed in public service in a single year in any river basin," said the *Journal*.

59. Other variations of the project (supported by Harry Morrison, and later, Senator Frank Church) were proposed through the 1970s, some advocating that groundwater replace Boise River water in the existing Boise Project. Established settlers feared that the scheme would disrupt their water supply and opposed these proposals. See Papers of Frank Church, Boise State University, "Guffey Project," Hillcrest Project," "Southwest Idaho Water Project." See also Scott W. Reed, "Dam Failure: The Swan Falls-Guffey Joint Venture (b. Jan. 1971 d. April 1979) R.I.P.," published in 1979 newsletters of Idaho Conservation League and Idaho Environmental Council. For additional discussion of Hells Canyon and the CVA, see Elmo Richardson, *Dams, Parks, and Politics: Resource Development and Preservation in the Truman-Eisenhower Era* (Lexington: University Press of Kentucky, 1973), especially Chapter One, "CVA: The Road Not Taken," p. 19–38.

60. *Statewide*, December 16, 1948, p. 4.

61. See "We Are Told We Have the Wrong Information, So We Correct an Opinion of Arrowrock Dam," *IDS*, August 16, 1953, p. 4, and "The Cost of a Bridge," *IDS*, July 1, 1952, p. 4.

62. "Aimed at Flood Control, Cost of Lucky Peak Dam Could Be Biggest Hoax; When Were the Floods?" *IDS*, August 2, 1953, p. 4.

63. See "Aimed at Flood Control, Costly Lucky Peak Dam Could Be Biggest Hoax; When Were the Floods?" *IDS*, August 2, 1953, p. 4. The Corps protested the false information and several other issues brought up in the editorial.

64. "Crowd Attends Dedication of Lucky Peak Project," *IDS*, June 24, 1955, p. 16.

65. "Wonder at Lucky Peak," *IDS*, June 25, 1955, p. 4. Other publications expressed similar views. *Statewide* editorialized on April 19, 1951, p. 4, "Why does

construction of Hells Canyon bring such anguished cries of bureaucracy and bungling, when the Lucky Peak dam above Boise was slipped through by Army Engineers without a murmur of protest? *Statewide* believes that the more than $20 million it will take to build Lucky Peak can never be justified."

## Chapter Four

1. Royse Van Curen, Project Manager for Boise Project Board of Control, "Historical Explanation Given on Flood Measures in Boise River's System," *IDS*, February 27, 1966, p. 7–B.

2. United States, Army Corps of Engineers, *Reservoir Regulation Manual for Boise River Reservoirs* (Walla Walla: District Engineers Office, August 1956), p. 16. (Hereafter referred to as *1956 Reservoir Regulation Manual.*)

3. United States, Corps of Engineers, *Minutes of Board of Consultants Meeting, June 7 and 8, 1949, Lucky Peak Dam, Boise River, Idaho* (Walla Walla: District Office). Appendix I of *Basis of Design, Definite Project Report on Lucky Peak Dam, Boise River, Idaho* is a summary of the minutes. Positive cut-off discussion is on p. I 6–7.

4. In an interview on June 25, 1987 at Walla Walla, Alvin Ross, Project Coordinator for Lucky Peak during its first years of construction, identified erosion prevention as the reason for the "rooster tail." However, the 1985 *Reservoir Regulation Manual* (p. 2–19) says that its main benefit is to prevent turbulence from endangering the integrity of the dam and outlet manifold structure. Dissipation of energy is accomplished "only to a small extent."

5. In practice, the Corps and Bureau tried to cooperate with downstream interests as much as possible whenever requests were made for short-term release adjustments. A typical example was a request by Boise City in 1957 to reduce flows to facilitate repair work at Lander Street Sewage Treatment Plant. Mayor Edlefsen to District Engineer, May 23, 1957, SRC, 77–78–0017, Box 30/49, "Water Flow."

6. "Memorandum of Agreement between the Department of the Army and the Department of the Interior for Flood Control Operation of Boise River Reservoirs, Idaho." Reprinted as "Appendix A" in *1956 Reservoir Regulation Manual*, p. A1–A14. (Hereafter referred to as "Memorandum.")

7. *1956 Reservoir Regulation Manual*, p. 19.

8. Robert Rickel, Chief, Hydrology Branch, Walla Walla District Office, telephone interview with author; and Dave Reese, Hydrology Branch, interview with author, June 25, 1987; also, Idaho Department of Water Resources, *Review of Boise River Flood Control Management* (Boise: Department of Water Resources, 1974), p. 49.

9. "Memorandum," p. A-13.

10. United States, Department of the Interior, Fish and Wildlife Service,

*Supplementary Follow-Up Report for Lucky Peak Dam Project, Idaho* (Portland: Office of the Commissioner, 1960), p. 9. (Hereafter referred to as *Follow-Up Report*.)

11. *Definite Project Report*, p. J-38.

12. *Follow-Up Report*, p. 9.

13. United States, Army Corps of Engineers, *Basis of Design, Definite Project Report on Lucky Peak Dam, Boise River, Idaho* Vol. 2 (Walla Walla: Office of the District Engineer, 1949), p. J-5. In 1986, Boise City's amended general plan also called for periodic "flushing flows" in the Boise River (not for malaria, but to help maintain the channel), but the Corps did not adopt the policy into the operating procedures. Boise City Ordinance 4863, September 30, 1985, Amendments to "A Policy Plan for the Boise Metropolitan Area," Flood Protection Goal 1d.

14. Paul Quick, Acting Director, Fish and Wildlife, to Colonel Mills, Walla Walla District, Corps of Engineers, 29 May 1951, SRC, 77–78–0017 Box 6, NPW 800.217 (Lucky Peak).

15. Colonel Mills to Paul Quick, July 10, 1951, SRC, 77–78–0017 Box 6, NPW 800.217 (Lucky Peak) 7.7.

16. Ross Leonard and J.R. Smead to Col. A.H. Miller, September 8, 1954, SRC, 77–78–0017 Box 6.

17. Leonard sent a letter to Colonel Miller to "express our appreciation for the cooperative attitude displayed by the Corps of Engineers," October 12, 1954. SRC, 77–78–0017, Box 5, NPW 337, "Lucky Peak."

18. Mark Kulp to Colonel A.H. Miller, October 18, 1954, SRC, 77–78–0017 Box 5, NPW 337, "Lucky Peak."

19. "Water Shutoff Arranged to Complete Dam Outlets," *IDS*, October 1, 1954, p. 3.

20. Leonard to Miller, October 12, 1954, SRC, 77–78–0017 Box 5 NPW 337.

21. Colonel A.H. Miller to Mark Kulp, 19 October 1954, SRC, 77–78–0017 Box 6 821.206.

22. Kulp to Miller, 24 November 1954, SRC, Npw 821.3 (Lucky Peak) 25 dm x 800.5.

23. According to the Fish and Wildlife *Follow-Up Report*, p. 10, there were no water releases from Lucky Peak during the following periods:

    November 1, 1954 to February 21, 1955    October 17, 1955 to January 18, 1956

    March 12, 1955 to March 24, 1955    October 14, 1956 to February 3, 1957

Flows of less than 1 cfs occurred in 1957 between January 8 and February 3, and between March 18 and March 22.

24. "Water Right Brief Hits Fish Rights," *IDS*, February 8, 1958, p. 1, and January 31, 1958, p. 1.

25. "Water Users Fight Game Group Claims of Fish Priority," *IDS*, February 1, 1958, p. 1.

26. Niel F. Meadowcroft, Hydraulic Engineer, notes on "State of Idaho Water Right Hearing Relative to Release of Water for Fish Purposes below Lucky Peak Dam," 12 February 58, SRC, 77–78–0017 Box 6 821.206.

27. "Appeal to Divert Water for Fish Turned Down," *IDS*, March 15, 1958, p. 1.

28. The Idaho Legislature passed a law in 1976 providing for fish habitat protection as a beneficial use of water.

29. P.L. 85–624. This 12 August 1958 Act amended the Fish and Wildlife Coordination Act of 14 August 1946, P.L. 732, 79th Cong.

30. "Lucky Peak Reservoir, Repayment of Irrigation Storage," Memorandum from North Pacific Division to Chief of Engineers, Walla Walla, 22 September 1965, SRC, 77–86–0006, 1520–03, Lucky Peak, Book 2.

31. "Fishery in Relation to Lucky Peak Water Releases, Meeting, Boise River: Corps of Engineers, Sportsmen, Fish and Game," June 6, 1957, p. 4, copy provided by Region 1, Bureau of Reclamation, Boise, Idaho.

32. Yaeckel to Sen. Henry Dworshak, April 26, 1956, SRC, 77–78–0017, Box 30. The area referred to was between Eagle and Parma on the Boise River.

33. "Flood Control, Boise River," September 27, 1957, SRC, 77–78–0017 Box 30, File NPW 800.5. The area of concern was the lower valley.

34. *Lucky Peak Dam and Reservoir* (Walla Walla: Corps of Engineers, 1955.)

35. District Engineer to Division Engineer, 2 April 1956, SRC, 77–78–0017 Box 30, Boise River, 123 bl.

36. Geo. Pickett to Division Engineer, SRC, 77–78–0017 Box 30, File 800.55 Flood Control.

37. In more recent years, the District Engineer did not have the same independence to spend funds earmarked for emergencies. The President must make a declaration of emergency first. Charles McKinley noted that in earlier decades, "The legal definition given by Congress to 'emergency' for the use of the emergency funds granted the chief engineer permits a generous interpretation of that term and opens the way to pressure-politics allocations. . . . The allocation of these funds is normally the result of local group demand upon the chief engineer, often buttressed by the local congressman's intervention." McKinley, p. 81. The Boise River records would support McKinley's observation. The chain of correspondence invariably begins with a letter to a senator or representative, folowed by a Congressional letter to the Corps, a request to Division from District office to approve fund expenditure, then approval, and finally a happy letter from Corps back to the senator or representative.

38. See *Downstream Channel Requirements*, p. 12–15.

39. *Downstream Channel Requirements*, p. 34.

40. *Downstream Channel Requirements*, p. 18. The Corps would have recom-

mended a capacity of 12,000 cfs except that the engineers felt the tunnel and canal from below Anderson Ranch Dam intended to take Boise River water to the Mountain Home desert would be carrying a constant flow of 1,900 cfs "in the foreseeable future," p. 32.

41. United States, Army Corps of Engineers, *Boise Valley Flood Control Project, Boise River,* Design Memorandum No. 2 (Boise: Cornell, Howland, Hayes and Merryfield Consulting Engineers, for Walla Walla District, 1963), Exhibit C. (Hereafter referred to as *Boise Valley Project.*)

42. *Boise Valley Project,* p. 38.

43. "Burns Issues New Warning to Sportsmen," *IDS,* September, 15, 1963, p. 5; and "C. Leo Holt Slaps Engineers On Boise River Dredge Plan," *IDS,* September 18, 1963, p. 6.

44. *IDS,* September 15, 1963, p. 5.

45. *IDS,* September 15, 1963, p. 5.

46. *IDS,* September 18, 1963. p. 6.

47. *Annual Report of the Chief of Engineers, 1967.* The summary memo of 17 August 67 in 1501–07, provides a chronology of Boise Valley Project.

48. Walla Walla District files, 1501–07, Boise Project, 25 July, 1969.

49. United States, Department of the Army, Corps of Engineers, Walla Walla District. *Levee Restudy on Boise River, Ada County, Idaho,* January 1976.

50. Idaho State Department of Water Resources, *Review of Boise River Flood Control Management.* (Boise: November, 1974.)

51. Dave Reese, interview with author in Walla Walla District Office, June 25, 1987.

52. Susan M. Stacy, *Population, Settlement, and Demographics* (Boise: Boise City Department of Community Planning and Development, 1984), p. 1.

53. The Clean Water Act of 1972 required that municipalities plan to meet current water quality standards.

54. Roy Ellerman, EPA, to Mayor Jay Amyx, Boise, April 28, 1971, Boise City Public Works Department, "Water Quality Standards for Flow."

55. Bill Ancell, Boise City Public Works Director, interview with author in Boise, August 28, 1986.

56. Corps engineers were also becoming increasingly concerned over the "blistering" problems in the one tunnel. On the inlet side, the part of the tunnel between the intake tower and the embankment was susceptible to buckling under pressure of the water above it when the tunnel was closed for inspection and repair. For this reason, the reservoir was drained to reduce the pressure before inspections were scheduled. It was an increasing inconvenience and another source of concern to irrigators.

57. OCE Regulation #1105–2–507 of April 15, 1974 entitled, "Preparation and Coordination of Environmental Impact Statements," p. 5. (Source office code:

DAEN–CWP–V.) See also Jeffrey Stine, "Environmental Politics and Water Resources Development: The Case of the Army Corps of Engineers During the 1970s." (Ph. D. dissertation, University of California at Santa Barbara, 1984); and also *Code of Federal Regulations*, "Protection of the Environment," 40, Section 1506.12 (b), "NEPA shall continue to be applicable to actions begun before Jan. 1, 1970, to the fullest extent possible." See also Frederick R. Anderson, *NEPA in the Courts: A Legal Analysis of the National Environmental Policy Act* (Baltimore: Johns Hopkins Press, 1973), particularly Chapter V, "Problems of Transition: Projects and Programs in Progress when NEPA was Enacted;" Richard A. Liroff, *A National Policy for the Environment: NEPA and its Aftermath* (Bloomington: Indiana University Press, 1976); and Beatrice Hort Holmes, *History of Federal Water Resources Programs and Policies* (U.S. Dept. of Agriculture, Economics, Statistics, and Cooperative Service Miscellaneous Publication No. 1379, 1979).

58. United States, Corps of Engineers, *Draft Environmental Impact Statement: Lucky Peak Dam and Lake, Boise River, Idaho* (Walla Walla: Office of the District Engineer, 1974), pages 1, 105.

59. William Ancell to Col. Conover, May 29, 1974, Boise City Public Works Department, "Lucky Peak."

60. Harold Geren, EPA, to William Ancell, June 26, 1974, Boise City Public Works Department, "West Boise Water Quality Standards for Flow."

61. Vic Armacost, Corps, to Engineering Division Files, 5 June 1974, SRC, 77, 86–0006, Box 1/3, Book 3.

62. United States, Corps of Engineers, *Lucky Peak Dam and Reservoir Modification Study Informational Brochure #2* (Walla Walla: Office of the District Engineer, November 1975) p. 4.

63. William Ancell, August 28, 1986.

64. Senator Frank Church to Mayor Dick Eardley, October 8, 1976, Boise City Public Works Department, "Lucky Peak." See also "Lucky Peak Dam Outlet Approved by Congress," *IDS*, October 5, 1976, p. B-6; and PL 94–587, 94th Cong., 2nd. Sess., 1976.

65. In addition, there would be little advantage to one of the traditional allies of the irrigators, the Idaho Power Company. The irrigators initially sought an arrangement with Idaho Power before taking up negotiations with other interests. See Susan M. Stacy, *Legacy of Light* (Boise: Idaho Power Company, 1991), p. 225.

66. "Lucky Peak Proposal Unveiled, Irrigation Districts Seek Power Plant," *IDS*, October 7, 1976, p. B-1.

67. John Jameson, Keith Petersen, Mary Reed, *Walla Walla District History, Part III, 1975–1980* (Walla Walla: Army Corps of Engineers District Office, 1980), p. 14.

68. Board of Control members, as they argued among themselves, entertained

competitive proposals from Idaho Power, J.R. Simplot, and others, and eventually entered into a deal with Seattle City Power and Light. The Boise Project Board of Control broke every promise it made in its 1976 statement. Power production is still a potential at Arrowrock, and the Board has the license; Board wrangling caused such delays that Lucky Peak construction did not begin until 1986; the board marketed 100% of the power not only outside Boise Valley, but outside the entire state of Idaho.

69. For general discussion of the Corps and the environmental movement, see Daniel Mazmanian and Jeanne Nienaber, *Can Organizations Change? Environmental Protection, Citizen Participation, and the Corps of Engineers* (Washington: Brookings Institution, 1979) and Jeffrey Stine, "Environmental Politics and Water Resources Development: The Case of the Army Corps of Engineers During the 1970s" (Ph.D. dissertation, University of California at Santa Barbara, 1984).

70. Dave Reese, June 25, 1987.

71. United States, Army Corps of Engineers, *Water Control Manual for Boise River Reservoirs* (Walla Walla: District Office, 1985), p. E, 7-1, 7-2. (Hereafter referred to as *1985 Reservoir Regulation Manual*.)

72. See *1985 Reservoir Regulation Manual*, p. 7-31 and 7-2.

73. Dave Reese, June 25, 1987.

74. *Idaho Business Review*, January 13, 1986, p. 1. After the debt is retired in 20 years, the districts expect to receive annual revenue of $12–15 million.

75. Dave Reese, June 25, 1987.

## Chapter Five

1. Charles Hummel, interview with author in Boise, July 18, 1986.

2. United States, Department of the Interior, Fish and Wildlife Service, *Supplementary Follow-Up Report for Lucky Peak Dam Project, Idaho* (Portland, Oregon: Dept. of the Interior, 1960), p.10. The closure began on November 1, 1954, and lasted until Febuary 21, 1955.

3. See William Webb, "The Boise River, A Problem Stream," *Idaho Wildlife Review* (September–October 1964), p. 10; William E. Webb, "General Investigation in Water Quality," *Fish and Game Publications* 11 (Boise: Idaho Fish and Game Department, 1962), p. 53; and United States, Corps of Engineers, *Basis of Design, Definite Project Report on Lucky Peak Dam, Boise River, Idaho*, Vol. 1 (Walla Walla: Office of the District Engineer, 1949), p. A-15.

4. Osborne E. Casey, "Water Quality Investigations," *Fish and Game Publications* 6 (Boise: Idaho Fish and Game Department, 1959), Report #72, p. 1.

5. William Webb, "The Boise River, A Problem Stream," *Idaho Wildlife Review* (September–October 1964), p. 10.

6. Jon Roten, Senior Water Quality Specialist, Division of Environment, Idaho

Department of Health and Welfare, telephone interview with author in Boise, May 18, 1987.

7. Eugene B. Chaffee, *Boise College, An Idea Grows* (Boise, 1970), p. 60–61. See photograph facing page 27. Chaffee said that it seemed logical to place the building at the center of the campus and facing the highest peak on the Boise Front, Shafer Butte. The line between the building and Shafer was also at right angles to a tangent of a curve in the Boise River.

8. Gordon Bowen, former Boise City Parks Director, interview with author in Boise, August 26, 1986.

9. Atkinson Associates, *Comprehensive General Plan, Boise City, Idaho, 1985* (Boise, 1963), p. 20. (Hereafter referred to as *Atkinson Report*.)

10. *Atkinson Report*, p. 35.

11. Gordon Bowen, August 26, 1986.

12. The resolution was adopted on January 24, 1966. Since this was in the middle of the budget year, no funds were expressly appropriated for the project at that time.

13. Alice Dieter, former member of Boise City Parks Commission, interview with author, in Boise, May 15, 1987.

14. Boise City Park Department, *1967 Annual Report*, (Boise: Park Department, 1968).

15. Arlo Nelson of Planning Research West, "Boise River Greenbelt Comprehensive Plan and Design" (Boise: Boise City Park Department, 1968).

16. Alice Dieter, "The Greenbelt: It's Happening," *Intermountain Observer* (February 22, 1969), p. 10.

17. See "Guidelines for Boise River Greenbelt," February 20, 1970; "Revised Guidelines as Amended;" and memorandum from Gordon Bowen to Mayor and City Council, July 23, 1970. Boise City Park Department file, "Rules, Regulations, and Guidelines."

18. United States, Corps of Engineers, *Flood Plain Information, Boise, Idaho and Vicinity; Boise River and Northside Tributaries* (Walla Walla: District Office, Corps of Engineers, 1967.) The authority for the report was Section 206, Flood Control Act of 1960, Public Law 86–645.

19. *Flood Plain Information*, p. i.

20. *Flood Plain Information*, p. 1.

21. *Flood Plain Information*, p. ii.

22. Nelson, p. 27.

23. Alice Dieter, *Intermountain Observer*, February 22, 1969, p. 10.

24. Public Law 92–500, Title II, Section 208 (h)(1). More specifically, the law authorized the Corps to provide technical assistance in "developing and operating a continuing areawide waste treatment planning process."

25. The results of the work were published as a series of Regional Water Quality Study Reports.

26. Boise City Park Department, *Annual Reports* for 1974, 1977. According to a summary report by the Park Director dated March 22, 1983, the projects included Main Street Tunnel, Fairview Underpass, Americana Boulevard Underpass, Broadway Underpass (South Bank).

27. Carrie Ewing, "Boise Editor Tells of Try to Correct `Death Trap,'" *IDS*, June 30, 1970, p. 4-C.

28. Boise City Park Department, *Annual Report for 1971*.

29. "Support for River Clean-Up Described to Chamber of Commerce Directors," *IDS*, October 19, 1971, p. 28; Ruth Russell, "Volunteers Hasten Boise River Clean Up," November 22, 1971, p. 1-D; Tim Woodward, "Boise River Hazards Still Found Despite Fall Clean-Up Campaign," July 12, 1972, p. 1-D.

30. Boise City Planning Department, File V–14–74.

31. Marsden and Associates, *Garden City Comprehensive Plan* (Garden City, Idaho, no date), p. O-23, O-24.

## Chapter Six

1. *Atkinson Report*, p. 10.

2. PL 90–448, 1 August 1968; Flood Insurance, Title XIII, (82 Stat. 572, 42 USC, 4001). See also Flood Disaster Protection Act of 1973, December 31, 1973, PL 93–234; (87 Stat. 975).

3. Boise entered the emergency program in April 1975, and the regular program in April 1984; Garden City in January 1979 and May 1980 respectively; Ada County in May 1975 and December 1984 respectively.

4. United States, Federal Emergency Management Agency, *Flood Insurance Study, City of Boise, Idaho, Ada County*, Report # 160002 (Bothell, Washington, 1984), p. 9. (Hereafter referred to as *Flood Insurance Study*.)

5. Flood Insurance Act of 1968, Public Law 90–448, Title XIII, (82 Stat. 572m 42 U.S.C. 4001), as amended; and Flood Disaster Protection Act, Public Law 93–234, (87 Stat. 975). See also *Flood Insurance Study*.

6. Federal Emergency Management Agency, *Flood Insurance Study, City of Boise, Idaho, Ada County* (FEMA: 1984), p. 4–9.

7. Kathleen Gager, interview with author in Boise, April 30, 1987.

8. Boise City Ordinance # 3924, December, 1976. The definitions cited in the footnote in the text are from the revision to the General Plan made in 1985, and are more technically accurate definitions of floodway and floodway fringe.

9. Annexation occurred on January 12, 1976. See Boise City Planning Department File A–12–75.

10. See Boise City Planning Department File CU–36–77.

11. Boise City Planning Department File CU–36–77.
12. See File CU–36–77, condition # 6 in letter of approval of March 22, 1977.
13. Presentation by Richard Holst to Select Committee on Downtown Boise, October 10, 1982, Boise, (author's notes of meeting.) Holst was project management consultant for Albertsons, Emkay, Ore-Ida, and other corporate clients in Boise.
14. Dick Eardley, former mayor, and Ralph McAdams, former city councilman, interviews with author in Boise, March 16, 1992, and March 21, 1992, respectively.
15. Minutes of the engineers' meetings and letters pertaining to them are in Boise Planning Department Files CU–36–77 and RZ–13–77 and Public Works Department file on ParkCenter Subdivision #1.
16. Record of meeting on March 29, 1977; presentation by Wendell Higgins, Boise City Public Works Department, File "ParkCenter #1."
17. Ron Barrett to Chuck Mickelson, November 23, 1977, Public Works Subdivision File 77–16–2, ParkCenter; Barrett to Mickelson, December 23, 1982, Walla Walla District file 1501–07, 1/3, ParkCenter.
18. Ron Barrett to Chuck Mickelson, December 23, 1982, Walla Walla District file 1501–07, 1/3, ParkCenter.
19. Ron Barrett to Chuck Mickelson, January 12, 1978, Walla Walla District Office File 82–25, "Certify ParkCenter." See also letter from Chuck Mickelson to Robert LeFevre, July 23, 1979, Boise Public Works department File ParkCenter #3. Mickelson warned the Emkay engineer that if the Loggers Creek diversion structure were not built as planned, the Federal Insurance Administration could change the official flood-plain boundaries and include all of ParkCenter subdivision #3 in the flood plain.
20. Gerry Unterkoefler, Director of Planning, to Ron Barrett, Corps, December 27, 1977, Walla Walla District File 78–33, "Loggers Creek."
21. Tony Peterson to Unterkoefler, March 14, 1978, Walla Walla District File 78–33, "Loggers Creek."
22. "River Run—A Request for Land Use Concept Approval," Boise City Planning Department File CU–34–78.
23. The approval condition read, "City must receive from Corps a letter and map delineating flood prone areas." Boise City Planning Department, File CU–34–78.
24. Boise City subsequently consummated the land trade in violation of federal rules and procedures.
25. To Peter O'Neill from Doug Erdman, March 24, 1978, File CU–34–78.
26. Boise City Planning Department, File CU–34–78. "Agreement with Greenbelt Committee," attachment to application.
27. Boise City Planning Department, CU–34–78, River Run application for concept plan, Exhibit D.

28. To Tony Peterson from Ron Barrett, April 20, 1978; Boise City Planning Dept. File A–8–78.

29. To Jack Cooper from Peter O'Neill, September 19, 1979, Boise City Planning Department, "Land Trade–River Run." See also explanation from Park Director Jack Cooper to the Greenbelt Committee on September 10, 1979. River Run used the fill from the auxiliary channel to construct its levee.

30. To Jack Cooper from Peter O'Neill, May 13, 1980, Attachment II. Park Department River Run file box.

31. River Run promotional brochure, "The Island." Undated, a copy is in the River Run files at the Planning Department.

32. To Jack Cooper from Peter O'Neill, September 19, 1979, Planning Department file, "River Run Land Trade."

33. Interview by author with Peter O'Neill, March 18, 1992, in Boise.

34. March 19, 1980, Greenbelt Committee Minutes; Park Department River Run file box.

35. City Council minutes, November 17, 1980, Resolution 6556, "Exchange Agreement."

36. Peter O'Neill to Mayor Dick Eardley, December 23, 1980. Boise Planning Department, file "River Run Land Exchange." In 1986, the park dirctor noted that this document had never been taken before the city council and "could not be considered a valid and binding instrument." See Cooper's notes in special River Run file in Park Department's River Run file box. A related issue was that River Run wanted the city to guarantee the delivery of a small sliver of land on the eastern end of the development known as the "Rose remnant." The city had obtained this parcel by means of Land and Water Conservation funds and had to obtain HCRS approval first, which approval had not been granted by December 1980. The whole issue of the questionable land trades came to light early in 1986 when River Run wished to plat one of its subdivision phases on the Rose remnant without yet owning it.

37. To Ed Miller from Jack Cooper, October 17, 1980; to Greenbelt Committee from Jack Cooper, May 13, 1983, Park Department River Run file box.

38. In 1990 the Urban Land Institute awarded River Run an "Award for Excellence for Large-Scale Residential Development," stating that the developer's "attention to detail and commitment to quality and the environment have created a premiere planned development that serves as an example for the entire region." The award was on display at Peter O'Neill's office in Boise.

39. Boise City Planning Department, File RZ–13–77, Minutes of Meeting, April 13, 1977.

40. Jim Floros, letter in December 22, 1986, editorial page, *Idaho Statesman*. See similar protests in the *Idaho Statesman*: December 24, 1986; March 11, 1987;

March 12, 1987; March 18, 1987; May 3, 1987; May 4, 1987; July 6, 1987; August 12, 1987; August 19, 1987 (editorial); April 4, 1988; April 6, 1988.

41. To Jim Floros from Jack Cooper, December 22, 1986; Park Department River Run file box.

42. Interview with Kathleen Gager, member of the Planning Department staff, held in Boise, April 30, 1987.

43. Boise City Ordinance # 4298, adopted in October 16, 1978, "A Policy Plan for the Boise Metropolitan Area" (Metro Plan), p. 34–35.

44. Boise City Planning Department, File RZ–3–79 and CU–182–78, "Forest River." See Condition # 15 in letter of approval.

45. See Boise City Planning Department, File CU–182–78, letter from Mike Preston to Stanley Posma, April 6, 1979.

46. Chris Korte to Mayor Richard Eardley, April 17, 1979; Chuck Mickelson to Mayor Eardley, 6–21–79. Copies in Planning Department, River Run file.

47. Boise City Planning Department, File RZ–14–82, Ron Barrett to Chuck Mickelson, August 25, 1982.

48. The Federal Water Pollution Control Act Amendments of 1972 (Public Law 92–500) in Section 404 required that the Corps of Engineers issue a permit for the disposal of fill or dredged material below the mean high water mark of rivers. The purpose is to prevent adverse impact on municipal water supplies, fisheries, and recreation.

49. Ron Barrett to Chuck Mickelson, 24 August 1982. Boise City Planning Department, File CU–44–84.

50. Councilman Glenn Selander made this remark at a Council work session.

51. Resource Systems, Inc., Karl Gebhardt, Project Manager, *Boise River Fish and Wildlife Habitat Study* (Boise City: Planning Department, 1983) p. 66.

52. Karl Gebhardt, et al., *Boise River Wildlife and Fish Habitat Study* (Boise: Boise City, 1983), p. 77. See also pages 71, 76, 90, 95, 102, 103, 105–106, 116–117.

53. To R.M. Chastain from John G. Hathaway, December 12, 1983, Walla Walla file 1501–07, 2/3.

54. To Susan Stacy from Ron Barrett, May 24, 1985; Walla Walla file 1501–07, 3/3.

55. Boise City Ordinance # 4863, "Boise River Plan," adopted September 30, 1985, p. 43–46.

56. Gebhardt, p. 41. See also page 75. The report attributes this comment to Karen Steenhof, a biologist for the Bureau of Land Management, and author of *Management of Wintering Bald Eagles*, U.S. Fish and Wildlife Service, Wash D.C., 1978.

57. *Garden City Comprehensive Plan*, Garden City, no date, p. O-24.

Notes to Pages 100–108

58. Randy Stapilus, "Plantation Fails to Meet Backers' Dreams," *IDS*, March 24, 1985, p. 1-A.
59. *IDS*, March 24, 1985, p. 1-A.
60. Ron Barrett to Roy Johnson, 1 December 1978, Walla Walla District Office, File 1501–07, Boise River, 1/3. See 404 permit Idaho # 0710 YC–4–000646.
61. "Proposed Route for Greenbelt Washed Away," *IDS*, May 6, 1984, p. 1-C.
62. Randy Stapilus, "Eroding Dreams," *IDS*, May 27, 1985, p. 1-A.
63. "Agreement," John V. Evans, Governor, and Riverside Group, (Nile Latta and Jack Hoke), November 10, 1980. Idaho Department of Lands. Copy in Boise City Planning Department "Area of Impact" file.
64. Susan Stacy to Mayor and Council, August 26, 1983. Boise City Planning Department file "Area of Impact." The author recommended that a recent request by Riverside Village for annexation to Boise City be considered seriously, in part so that "one of the four available river frontages are reserved for the continuation of a public access greenbelt without going to a street."
65. See "Riverside Village," *IDS*, May 17, 1987, p. 3-E.
66. Margaret Mockwitz to Mayor and City Council of Boise, February 24, 1983, Walla Walla file 1501–07.
67. Boise City Planning Department file "Area of Impact."

## Chapter Seven

1. Idaho Department of Water Resources, *Review of Boise River Flood Control Management* (Boise: November 1974), Table 3.
2. Larry Swisher, "High Water Poses Threat to Idaho Fun Seekers," *IDS*, May 28, 1983, p. 1-A.
3. *IDS*, May 29, 1983, p. 1-A.
4. "Floodwaters Flow as Teton River Bank Collapses," *IDS*, June 1, 1983, p. 1-C; "Boise River to Rise 3 More Inches," June 8, 1983, p. 1-A.
5. Jack Blake, interview by author in Boise, June 15, 1987; "Flow Bulges to Record Level," *IDS*, June 2, 1983, p. 1-A.
6. "River Bank Residents Evacuate Farm Animals," *IDS*, June 7, 1983, p. 1-A.
7. Ron Zellar,"It's 50-50 Chance the Boise Will Rise," *IDS*, June 7, 1983, p. 1-A.
8. "It's 50-50 Chance the Boise Will Rise," *IDS*, June 7, 1983, p. 1-A.
9. "50-50 Chance," *Idaho Statesman*, June 9, 1983, p. 1-A.
10. Interviews by author with Dave Reese in Walla Walla, June 25, 1987, and with Jack Blake in Boise, June 15, 1987. Also see *IDS*, June 12, 1983, p. 1-A.
11. "Boise River to Rise 3 More Inches," *IDS*, June 8, 1983, p. 1-A.
12. According to the *Flood Insurance Study for City of Boise, Idaho, Ada County* published by the Federal Emergency Management Agency, 1984, p. 8, "The hydraulic analyses for this study were based only on the effects of unobstructed

flow. The flood elevations, as shown in the profiles, are considered valid only if hydraulic structures remain unobstructed."

13. Interviews by author in Boise with Chuck Mickelson on March 11, 1987; Jack Blake on June 15, 1987; and Charles Winder on February 7, 1993. Also, "River to Rise; Floodwatch Intensifies," *IDS*, June 11, 1983, p. 1-A.

14. Ron Zellar and Ray Sotero, "River to Rise; Floodwatch Intensifies," *IDS*, June 11, 1983, p. 1-A.

15. Ron Zellar, "Floodway Violations Suspected, Sandbagging Dams, Channels," *IDS*, June 23, 1983, p. 1-A.

16. "Boise River to Rise 3 More Inches," *IDS*, June 8, 1983, p. 1-A.

17. To Jack Blake from Jack Hansen, June 9, 1983, office of Ada County Civil Defense Coordinator.

18. "Swollen River to Rise Some More," *IDS*, June 9, 1983, p. 1-A.

19. "The Relentless River," *IDS*, June 10, 1983, p. 1-A.

20. "River to Rise, Floodwatch Intensifies," *IDS*, June 11, 1983, p. 1-A, 1-C.

21. "Coolness Causes River-Boost Delay," *IDS*, June 12, 1983, p. 1-A; "Runoff Drops, May Equal Flow Out of Dams," *IDS*, June 14, 1983, p. 1-A.

22. Table provided by Dave Reese, Walla Walla District Office, "Lucky Peak—Total Unregulated Inflow (Kcfs), 1 October 1982 thru 30 September 1983."

23. "Sandbags Belie Floodway Promises," *IDS*, June 26, 1983, p. 2-D.

24. Ron Zellar, "Floodway Violations Suspected; Sandbagging Dams, Channels," *IDS*, June 23, 1983.

25. John E. Blake, *Ada County Hazard Vulnerability Analysis, 1985* (Boise: Ada County Civil Defense, 1985), p. 16.

26. The 1974 discharge-damage study was made before most of the new riverfront development. The gap between the half-million dollar estimate in 1974 and the lower actuals for a flood only 500 cfs less would seem to place the reliability of the discharge-damage charts in some doubt. See *Levee Restudy*, p.10. During the course of the 1983 flood, the Corps public affairs officer, O.C. Duggar, announced on two occasions what amount of damage could be expected at subsequent releases: at 8,000 cfs, he said the damage would be worth $500,000; at 9,000 cfs, $1,500,000; at 10,000 cfs, $2,700,000. See *IDS*, June 8, 1983, p. 1-A and June 9, 1983, p. 1-A. He used 1983 dollar values, but the projections still provided for $1.2 million of additional damage if a flood escalated from 9,000 to 10,000 cfs. Since the flood peak was 9,500 cfs and no subsequent analysis of actual damage costs, it is not likely that the community will ever know how much the flood cost, how close the Corps projections were, what accounts for the gap between Jack Blake's report and the Corps estimates, or what level of damage to expect if the river ever flows at 10,000 cfs.

27. Debby Abe, "Floating Through Boise's Jungle; Watery Obstacle Course Prompts Clean-Up Drive," *IDS*, August 11, 1983, p. 1-A.

28. Personal communication from Jack Cooper to author, August 31, 1987.

29. "Sandbags Belie Floodway Promises," *IDS*, June 26, 1983, p. 2-D. The editor of the *Statesman* thought that flood levels had been increased from six to 12 inches in places where sandbags had been used. While no one had reported damage, the editorial suggested "we can't press our luck."

30. In a general discussion of available methods for reduction of flood damage, the Corps of Engineers in the Baltimore District wrote, "The construction of levee and floodwall projects may also worsen flood conditions at other locations upstream, downstream, and across the stream . . . may also destroy wildlife habitat . . . only a limited degree of flood protection can be provided." *Flood Damage Reduction Manual*, U.S. Army Corps of Engineers, Baltimore District, DP 500–1–80, May 1984, p. 10.

31. *IDS*, June 12, 1983, p. 3-D.

32. Interview by author with Dave Reese in Walla Walla on June 25, 1987.

33. For an economic analysis of the benefits and costs of Lucky Peak in comparison to those which justified the project, see Yoseph Gutema, "An Expost Study of the Economic Performance of Federal Investments in Flood Control Projects in the Boise Valley, Idaho," Masters Thesis, University of Idaho, 1977.

## Chapter Eight

1. Ernst Breisach, *Historiography: Ancient, Medieval, and Modern* (Chicago: University of Chicago Press, 1983), p. 18.

2. Chuck Mickelson, interview with author in Boise, March 11, 1987.

3. *Metro Plan*, 1985, p. 43. See also: Bob Rickel to Kathy Gager, December 1, 1983, Walla Walla District file 1501–07 2/3.

4. Dave Reese, interview with author in Walla Walla, July 11, 1987.

5. Chuck Mickelson, interview with author in Boise, March 11, 1987.

6. Arthur Maass, *Muddy Waters* (Cambridge: Harvard University Press, 1951), p. 3.

7. Maass, p. 4.

8. Maass, p. 207. Many others followed Maass in criticizing the Corps of Engineers, accusing it of a straight-line engineering mentality, arrogance, isolation, or careless attitudes toward the environment—as though such evils did not exist in any other sector of American society. See as an extreme polemical work Gene Marine's *America the Raped, The Engineering Mentality and the Devastation of a Continent* (New York: Simon and Schuster, 1969). Dr. Todd Shallat of Boise State University surveyed the titles in the bibliography *United States Army Corps of Engineers and Environmental Issues in the Twentieth Century* by Jeffrey K. Stine and Michael Robinson (Wash DC: Historical Division, Office of the Chief of Engineers, 1984) and found 160 which expressed an openly hostile attitude to the

Corps. The words, "devastation," "disaster," or "destruction" appeared 19 times, "pork barrel," 14 times, and "boondoggle," 10 times.

9. Local interests in Boise were part of a national consensus. On May 17, 1943, the *Idaho Statesman* carried a page one story about an address made by Vice President Henry A. Wallace to the American Labor Party in which he said that the post-war plan being drafted by the Natural Resources Planning Board was to offset the "economic shock" when peace returns in case "private employment is not adequate to face the shock alone." To "preserve the dynamic character of our economy" was more important than dealing with unemployment, he said.

10. John Ferejohn, *Pork Barrel Politics: Rivers and Harbors Legislation, 1947–1968* (Stanford: Stanford University Press, 1974), p. 5.

11. Leonard Shabman, "Non-Market Valuation and Public Policy: Historical Lessons and New Directions." Paper prepared for Southern Natural Resource Economics Committee Meeting, Biloxi, Mississippi, May 19, 1983.

12. See Daniel Mazmanian and Jeanne Nienaber. The authors suggest that the Corps changed and willingly, p. 63. Also, see Jeffrey Stine, "Environmental Politics and Water Resource Development: The Case of the Army Corps of Engineers During the 1970s," (Ph.D. dissertation, University of California, Santa Barbara, 1984). Stine argues that change in the Corps came as a result of massive environmental litigation. Serge Taylor, in *Making Bureaucracies Think, The Environmental Impact Statement Strategy of Administrative Reform* (Stanford: Stanford University Press, 1984), evaluates whether the standards of scientific analysis fare better in the political decision environment as a result of the EIS process.

13. Personal communication from Thomas Slater, responsible for managing the Walla Walla District EIS program for Lucky Peak, December 1, 1988.

14. See United States Department of Agriculture, *Survey Report: Boise River Watershed* (Boise: 1950).

15. "The Relentless River," *IDS*, June 13, 1983, p. 1-A.

16. Everett Rising, Address to National and Regional Land and Water Organizations meeting, September 18, 1947, printed in Everett Rising Newsletter #7, November 7, 1947, Papers of Henry Dworshak, IHS, Box 17, "Misc.–Rising, E.W.–1947."

17. Rochelle L. Stanfield, "A New Era," *National Journal* 47 (November 22, 1986), p. 2824.

# Bibliography

### Books and Journals

Achenbaum, W. Andrew. "The Making of an Applied Historian: Stage Two." *Public Historian* 5 (Spring 1983): 21–46.

Andrews, Richard N. L. "Environment and Bureaucracy: Progress and Prognosis." *Journal of Environmental Education* 6, no. 1 (Fall 1974): 1–6.

Arrington, Leonard J. "Idaho and the Great Depression." *Idaho Yesterdays* 13 (Summer, 1969): 2–8.

Banfield, Edward C. "Policy Science as Metaphysical Madness." In *Bureaucrats, Policy Analysts, Statesmen: Who Leads?* Edited by Robert A. Goldwin. Washington, DC: American Enterprise Institute for Public Policy Research, 1980

Barrett, Glen. "Reclamation's New Deal for Heavy Construction, M-K in the Great Depression." *Idaho Yesterdays* 22 (Fall 1978): 21–27.

Blackwelder, Brent. "In Lieu of Dams." *Water Spectrum* 9 (Fall 1977): 41–46.

Breisach, Ernst. *Historiography: Ancient, Medieval, and Modern.* Chicago: University of Chicago Press, 1983.

Burby, Raymond J., and Steven P. French. "Coping with Floods, the Land Use Management Paradox." *Journal of the American Planning Association* 47 (July 1981): 289–300.

Chaffee, Eugene B. *Boise College, An Idea Grows*. Boise, 1970.

———. "Boise, the Founding of a City." *Idaho Yesterdays* 7 (Summer, 1963): 2–7.

Chandler, John, and Arthur Doyle. "An Alliance With Nature." *Water Spectrum* 10 (Summer 1978): 24–31.

Clark, D. Worth. "Idaho Made the Desert Bloom." *National Geographic* 85 (June 1944): 641–88.

Drew, Elizabeth B. "Dam Outrage: The Story of the Army Engineers." In *The Military and American Society: Essays and Readings*. Edited by Stephen E. Ambrose and James A. Barker, Jr. New York: Free Press, 1972.

Donaldson, Thomas. *Idaho of Yesterday*. Caldwell: Caxton Printers, 1941. Reprinted by Greenwood Press of Westport, Conn., 1970.

Ferejohn, John. *Pork Barrel Politics: Rivers and Harbors Legislation, 1947–1968*. Stanford: Stanford University Press, 1974.

Fite, Gilbert. *The Farmers Frontier, 1865–1900*. New York: Holt, Rinehart, and Winston, 1966.

Florman, Samuel C. "Hired Scapegoats: In Support of the U.S. Army Corps of Engineers." *Harper's Magazine* 254 (May 1977): 26–29.

Gilliam, Harold. "The Engineering Mentality." *Natural History* 78 (Oct 1969): 72–73, 75–76. (Review of *America the Raped* by Gene Marine.)

Hill, Forest G. *Roads, Rails and Waterways, The Army Engineers and Early Transportation*. Norman, Oklahoma: University of Oklahoma Press, 1957.

Hogue, Gilber H. *Boise Project History, History of the Payette-Boise Project, Idaho*, Volumes 1 and 2. Boise, Idaho, March 1916. (Idaho Historical Library, MS 38).

Kingdon, John W. *Agendas, Alternatives, and Public Policies*. Boston: Little, Brown, and Co., 1984.

Koch, Stuart G. *Water Resources Planning in New England*. Hanover, New Hampshire: University Press of New England, 1980.

Lamar, Howard R., ed. *The Reader's Encyclopedia of the American West*. New York: Thomas Y. Crowell Co., 1977.

Lee, Lawrence B. *Reclaiming the Arid West, An Historiography and Guide*. Santa Barbara, California: ABC-Clio, 1980.

Lovin, Hugh T. "'Duty of Water' in Idaho, A 'New West' Irrigation Controversy, 1890–1920." *Arizona and the West* 23 (Spring 1981), p. 5–28.

———. "Free Enterprise and Large Scale Reclamation on the Twin Falls-North Side Tract, 1907-1930." *Idaho Yesterdays* 29 (Spring 1985): 2–14.

Lowell, Helen, and Peterson, Lucille. *Our First Hundred Years: A Biography of Lower Boise Valley 1814-1914*. Caldwell, Idaho: Caxton Printers, 1976.

Maass, Arthur A. "Congress and Water Resources." In *Bureaucratic Power in National Politics*, pp. 139–152. Edited by Francis E. Rourke. Boston: Little, Brown, and Co., 1972.

———. *Muddy Waters*. Cambridge: Harvard University Press, 1951.

McKinley, Charles. *Uncle Sam and the Pacific Northwest*. Berkeley: University of California Press, 1952.

Malone, Michael P. *C. Ben Ross and the New Deal in Idaho*. Seattle: University of Washington Press, 1970.

Mann, Dean E., and Helen M. Ingram. "Policy Issues in the Natural Environment." In *Public Policy and the Natural Environment*, p. 15–45. Greenwich, Conn.: JAI Press, Inc., 1985.

Marine, Gene. *America the Raped, The Engineering Mentality and the Devastation of a Continent*. New York: Simon and Schuster, 1969.

Mazmanian, Daniel, and Jeanne Nienaber. *Can Organizations Change? Environmental Protection, Citizen Participation, and the Corps of Engineers*. Washington: Brookings Institution, 1979.

Miller, James Nathan. "A Plan to Use our Floods, Not Fight Them." *Readers Digest* 102 (March 1973): 21–22, 24–26.

Moore, Jamie W., and Dorothy P. Moore. *The Army Corps of Engineers and the Evolution of Federal Flood Plain Management Policy*. University of Colorado: Natural Hazards Research and Applications Information Center, 1989.

Morgan, Arthur E. *Dams and Other Disasters, A Century of the Army Corps of Engineers in Civil Works*. Boston: Porter Sargent Publisher, 1971.

Muckleston, Keith W. "The Evolution of Approaches to Flood Damage Reduction." *Journal of Soil and Water Conservation* 31 (Mar-Apr 1976): 53–59.

Murphy, Paul L. "Early Irrigation in Boise Valley." *Pacific Northwest Quarterly* 44 (1953): 177–184.

Newell, R. J. "Water for the West." *Idaho Yesterdays* 2 (Spring 1958): 16–21.

Peterson, F. Ross, and W. Darrell Gertsch. "The Creation of Idaho's Lifeblood: The Politics of Irrigation." *Rendezvous* 11 (Fall 1976): 53–61.

Public Works Historical Society. *Public Works History in the United States*. Ed. Snellen M. Hoy, Michael C. Robinson, and Rita Lynch. Nashville: American Association for State and Local History, 1982.

Reuss, Martin. "Andrew A. Humphreys and the Development of Hydraulic Engineering: Politics and Technology in the Army Corps of Engineers." *Technology and Culture* 26 (January 1985).

Richardson, Elmo. *Dams, Parks and Politics, Resource Development and Preservation in the Truman-Eisenhower Era*. Lexington, Kentucky: University Press of Kentucky, 1973.

Rostow, W. W. *The World Economy: History and Prospect*. Austin, Texas: University of Texas Press, 1978.

Rourke, Francis E. *Bureaucracy, Politics, and Public Policy*. Boston: Little, Brown, and Co., 1984.

Notardonato, Frank. "Corps Takes New Approach to Flood Control." *Civil Engineering* 49 (June 1979), p. 65–68.

Reed, Scott W. "Dam Failure: The Swan Falls-Guffey Joint Venture (b. Jan 1971 d. April 1979) R.I.P." Published in Idaho Conservation League and Idaho Environmental Council newsletters in 1979.

———. "New Law for a New State, The Legal Impetus to Development of the Material Resources of Idaho." *Idaho Yesterdays* 25 (Spring 1981): 47–56.

———. "The Other Uses for Water." *Idaho Yesterdays* 30 (Spring/Summer, 1986): 33–44.

Schlesinger, Jr., Arthur M. *The Cycles of American History.* Boston: Houghton Mifflin, 1986.

Shabman, Leonard. "Non-market Valuation and Public Policy: Historical Lessons and New Directions." Paper prepared for Southern Natural Resource Economics Committee meeting, Biloxi, Mississippi, May 19, 1983.

Stanfield, Rochelle L. "A New Era." *National Journal* No. 47 (November 22, 1986): 2822–2825.

Taylor, Serge. *Making Bureaucracies Think, The Environmental Impact Statement Strategy of Administrative Reform.* Stanford: Stanford University Press, 1984.

Ullman, Edward L. "Rivers as Regional Bonds: The Columbia-Snake Example." *Geographic Review* 41 (April 1951): 210–225.

Warnick, C.C., and Brockway. *Hydrology Support Study, Boise Project.* Moscow, Idaho: University of Idaho, Water Resources Research Institute, June 1974.

Webb, William. "The Boise River, A Problem Stream." *Wildlife Review*, Sept.–Oct. 1964, pp. 8–11.

Wells, Merle. *Boise, An Illustrated History.* Woodland Hills, California: Windsor Publications, Inc., 1982.

———. "Boise River. High-Water Years of the Past." Reference Series # 879. Boise: Idaho Historical Society, 1987.

———. "Twenty Years of History." *Idaho Yesterdays* 20 (Winter 1977): 12–15.

White, Gilbert F. "A Perspective of River Basin Development." *Journal of Law and Contemporary Problems* 22 (1957): 157–87.

Williams, Kenneth J. "Sugar Beet Growing in Ada and Canyon Counties, Idaho." *Pacific Northwest Quarterly* 42 (July 1951): 203–210.

Worster, Donald. "New West, True West: Interpreting The Region's History." *The Western Historical Quarterly* 18 (April 1987): 141–156.

———. *Rivers of Empire: Water, Aridity, and the Growth of the American West.* New York: Pantheon, 1985.

Wykstra, R.A., and R. D. Peterson. "Economic Growth in Idaho from 1948-1964." *University of Washington Business Review* 26 (Spring 1967): 29–43.

Zeldin, Marvin. "Corps New Look in Flood Control: No Dams, Levees." *Audubon* 77 (July 1975): 103–104.

## Government Documents and Reports

*Local*

Ada Council Of Governments, Canyon Development Council, and United States Corps of Engineers. *Public Information Brochure on the Barber Dam Problem in Boise River near Boise, Idaho.* Walla Walla: Corps of Engineers 1974.

Atkinson Associates. *Comprehensive General Plan, Boise City, Idaho, 1985.* Boise City, 1963.

Blake, John E. *Ada County Hazard Vulnerability Analysis, 1985.* Boise: Ada County Civil Defense, 1985.

Boise City. *A Policy Plan for the Boise Metropolitan Area.* Boise City, 1985.

Garden City, Idaho. *Garden City Comprehensive Plan.* Garden City, Idaho, no date.

Resource Systems, Inc., Karl Gebhardt, Project Manager. *Boise River Fish and Wildlife Habitat Study.* Boise City: Planning Department, 1983.

## State

Cochnauer, Tim, and Bill Horton. "Reference Workbook for Use in Determining Stream Resource Maintenance Flows in the State of Idaho." *State of Idaho Fish and Game Department Fish Division Publications.* Volume 58, 1985, Article 10.

Idaho. Water Resources Board. *State of Idaho Interim Water Plan, Preliminary Report.* Boise, Idaho, 1972.

Idaho. Department of Water Resources. *Review of Boise River Flood Control Management.* Boise, Idaho: November, 1974.

Thomas, C.A. and N.P. Dion. *Characteristics of Streamflow and Groundwater Conditions in the Boise River Valley, Idaho.* U.S. Geological Survey Water Resources Investigations 38–74. Boise Idaho, December 1974. (Prepared in cooperation with Idaho Department of Water Resources and Walla Walla District Corps of Engineers.)

Idaho. Department of Water Resources. *Review of Boise River Flood Control Management.* Boise, Idaho, November 1974.

## Federal

*Barber Dam Advance Report: A Report of the Present Status of the Barber Dam Problem and Possible Solutions.* Walla Walla, U.S. Army Corps of Engineers, 1975. Regional Water Management Study. (With Ada Council of Governments and the Canyon Development Council.)

Caldwell, H.H., and Wells, M. *Economic and Ecological History Support Study, A Case Study of Federal Expenditures on a Water and Related Land Resources Project: Boise Project, Idaho and Oregon.* Idaho Water Resources Board with Idaho Water Resources Research Institute, June 1974. OWRT Title II, Project C–4202.

Chen, Carl W., and Wells, John T. *Regional Water Management Study, Boise Valley.* Lafayette, California, 1975, unpublished. Prepared for Walla Walla Corps of Engineers, Idaho Water Resources Board, and Idaho Department of Environmental and Community Services.

Commission on Organization of the Executive Branch of the Government. *Task Force Report on Water Resources and Power*, Vol. 2, June 1955.

*The History of the North Pacific Division, US Army Corps of Engineers, 1888 to 1965*. Portland, Oregon: North Pacific Division, 1969.

Jameson, John R., Kieth Petersen, and Reed, Mary E. *Walla Walla District History, Part III 1975–1980*. Walla Walla: U.S. Army Corps of Engineers, 1980.

Holmes, Beatrice Hort. *A History of Federal Water Resources Programs, 1800–1960*. Washington, DC: U.S. Dept. of Agriculture Economic Research Service, 1972.

[Preston, Howard] *A History of the Walla Walla District 1948–1970*. Walla Walla: U.S. Army Corps of Engineers, [1970].

Preston, Howard A. *Walla Walla District History, Part II 1970–1975*. Walla Walla: U.S. Army Corps of Engineers, 1976.

Reuss, Martin. *Water Resources People and Issues, An Interview with William R. Gianelli*. Washington, DC: U.S. Army Corps of Engineers Historical Division, 1983.

Stine, Jeffrey K., and Michael C. Robinson. *The US Army Corps of Engineers and Environmental Issues in the Twentieth Century*. Washington: Historical Division, Office of Administrative Services, Office of the Chief of Engineers, 1984.

United States. Bureau of the Census. *Fourteenth Census of the United States*, 1920. Washington, DC: 1921.

United States. Bureau of the Census. *Sixteenth Census of the United States*, 1940. Washington, DC: 1941.

United States. *Code of Federal Regulations*, "Protection of the Environment," 40.

United States. Congress. House. *Columbia River and Minor Tributaries*. H. Doc. 103, 73rd Congress, 1st Session, 1932.

United States. Congress. House. *The Columbia River: Supplemental Reports on the Bitterroot Valley, Cambridge Bench, Canby, Council, Crooked River, The Dalles (West Unit), Hells Canyon, Mountain Home (Payette Unit), Upper Star Valley, and Vale (Bully Creek Extension) Projects*, House Document 473, 81st Congress, 2nd. Session, Vol. 2.

United States. Congress. House. *Columbia River and Tributaries Northwestern United States*. H. Doc. 531, 81st Congress, 2nd session, 1950.

United States. Congress. House. *Snake River and Tributaries*. H. Doc. 190, 73rd Congress, 2nd Session, 1932.

United States. Department of Agriculture. Intermountain Forest and Range Experiment Station in Cooperation with Soil Conservation Service. *Appendix Survey Report Boise River Watershed, Idaho and Oregon*. 1949. (Draft.)

United States. Department of Agriculture. Field Flood Coordinating Committee No. 17B. *Survey Report, The Boise River*, in three parts. 1940. (Mimeographed.)

United States. Department of Agriculture. *Survey Report Boise River Watershed*. Idaho, 1950.

United States. Department of the Army. Corps of Engineers. *The Annual Reports of the Chief of Engineers*. 1940–1985.

United States. Department of the Army. Corps of Engineers, Office of the Chief of Engineers. *Digest of Water Resources Policies and Authorities*. EP 1165–2–1. Washington, 1983.

United States. Department of the Army. *Tributaries of Boise River, Vicinity of Boise, Idaho*. House Document #486, 89th Congress, 2nd Session. Letter from the Secretary of the Army . . . March 10, 1960. United States Government Printing Office.

United States. Department of the Army. Corps of Engineers. *Seminar Proceedings, Implementation of Nonstructural Measures*. Policy Study 83–G502, July 1983.

United States. Army Corps of Engineers, North Pacific Division. *Plan of Survey, Columbia River and Tributaries Review Study*, (Draft). Portland: Corps of Engineers, December 1971.

United States. Department of the Army. Corps of Engineers, Portland District. *Appendix B, Hearing, Review of Survey Report, Boise River, Idaho, With A View to Control of Floods. Proceedings of Hearing 14 Dec 1944.* Portland, 2 Jan 1946.

United States. Department of the Army, Corps of Engineers, Portland District. *Review of Survey Report, Boise River, Idaho, With A View to Control of Floods.* Portland, 2 Jan 1946.

United States. Army Corps of Engineers. *Basis of Design, Definite Project Report on Lucky Peak Dam, Boise River, Idaho* Vol 2 (Walla Walla: Office of the District Engineer, 1949).

United States. Army Corps of Engineers, Walla Walla District. *Boise Valley Flood Control Project, Boise River.* Design Memorandum No. 2. Boise: Cornell, Howland, Hayes and Merryfield, June 1963.

United States. Corps of Engineers. Walla Walla District. *Boise Valley Regional Water Management Study, Summary Report.* Walla Walla, Washington, July 1977.

United States. Department of the Army. Corps of Engineers, Walla Walla District. *Definite Project Report on Lucky Peak Dam, Boise River, Idaho.* Walla Walla, October 3, 1949.

United States. US Army Corps of Engineers, Walla Walla District. *Downstream Channel Requirements, Lucky Peak Dam and Reservoir, Boise River, Idaho.* Design Memorandum No. 6. Walla Walla, September 1958.

United States. Army Corps of Engineers, Walla Walla District. *Draft Environmental Impact Statement: Lucky Peak Dam and Lake, Boise River, Idaho.* Walla Walla, March 1974.

United States. Army Corps of Engineers, Walla Walla District. *Final Environmental Impact Statement, Lucky Peak Dam and Lake, Boise River, Idaho.* Walla Walla, August 1976.

United States. Department of the Army. Corps of Engineers, Walla Walla District. *Flood Plain Information, Boise, Idaho and Vicinity; Boise River and Northside Tributaries.* Walla Walla, 1967.

United States. Department of the Army, Corps of Engineers, Walla Walla District. *Levee Restudy on Boise River, Ada County.* Idaho, Jan 1976.

United States. Department of the Army. Corps of Engineers, Walla Walla District. *Lucky Peak Dam and Reservoir 1955.* Informational Brochure.

United States. Department of the Army. Corps of Engineers, Walla Walla District. *Lucky Peak Dam and Reservoir Flow Maintenance Study,* Information Brochure #1 for Public Meeting 17 October 1974. Walla Walla, 1974.

United States. Department of the Army. Corps of Engineers, Walla Walla District. *Lucky Peak Dam and Reservoir Modification Study (CRT).* Information Brochure #2 for Public Workshop November 1975. Walla Walla, 1975.

United States. Department of the Army. Corps of Engineers, Walla Walla District. *Lucky Peak Dam and Reservoir Flow Maintenance Study Record of Public Meeting,* Boise, Idaho, 17 October 1974. (CRT). Walla Walla, Jan 1975.

United States. Department of the Army. Corps of Engineers, Walla Walla District. Design Memorandum No. 5: *The Master Plan for Development and Management of Lucky Peak Reservoir, Volume 1—Main Report and Appendices A and B.* Portland 1955.

United States. Department of the Army. Board of Engineers for Rivers and Harbors, *Survey Report of Boise River,* 76th Cong., 3rd sess., H.Doc. 957, Sept. 23, 1940.

United States. Department of the Army. Corps of Engineers. *Minutes of Board of Consultants Meeting, June 7 and 8, 1949, Lucky Peak Dam, Boise River, Idaho.* Walla Walla, District Office, 1949.

United States. Corps of Engineers, Walla Walla District; Ada Council of Governments, Canyon Development Council. *Plan of Study, Boise Valley Idaho Regional Water Management Study.* Walla Walla, Washington, June 1973.

United States. Department of the Army. Corps of Engineers, Walla Walla District. *Reservoir Regulation Manual for Boise River Reservoirs.* Walla Walla: Office of the District Engineer, August 1956.

United States. Department of the Army. Corps of Engineers, Walla Walla District. *Upper Snake River Basin Interim Report No.6, Lucky Peak Power Plant, Volume II.* Walla Walla, 1968.

United States. Department of the Army. Corps of Engineers, Walla Walla District. *Water Control Manual for Boise River Reservoirs.* Walla Walla, 1985.

United States. Department of the Interior. Bureau of Reclamation, Region I. *Scriver Creek Power Facilities, First Stage of Development of Power Features of the Overall Mountain Home Development.* Boise, Feb. 1951.

United States. Department of the Interior. Bureau of Reclamation, Region I. *Payette Unit Mountain Home Project, Idaho, A Supplement to the Columbia River Basin Report.* Boise, May 1949.

United States. Department of the Interior. Bureau of Reclamation. "Initial Unit, Mountain Home Project, Idaho," Report #69–A.

United States. Department of the Interior. Bureau of Reclamation, Region 1. *Snake River Project, Idaho, Oregon Composed of Mountain Home Division and Garden Valley Division, Report Summary.* Boise, May 1965.

United States. Department of the Interior. Fish and Wildlife Service Office of Regional Director, Region 1. *Lucky Peak Dam Project, Boise River, Idaho, Preliminary Evaluation Report on Fish and Wildlife Resources.* Portland, April 1950.

United States. Department of the Interior. Fish and Wildlife Service. *Supplementary Follow-Up Report for Lucky Peak Dam Project, Idaho.* Arnie J. Suomela, Commissioner. Portland, Jan 1960.

United States. Department of the Interior, Geological Survey. Thomas, Cecil Albert. *Cloudburst Floods at Boise, Idaho, August 20, September 22, 26, 1959.* Boise, Idaho, 1963.

United States. Federal Emergency Management Agency. *Flood Insurance Study, City of Boise, Idaho, Ada County*, Report # 160002. Bothell, Washington, 1984.

Waldorf, Dr. Larry, Dr. Richard K. Hart, and Donna Parsons. *Current and Projected Recreational Demand on the Lower Boise River.* Walla Walla: U.S. Corps of Engineers, May 1975. Contract #DACW 68–74–C–0188. A Publication of the Center for Business and Economic Research, Boise State University.

## Theses and Dissertations

Carlton, Neil H. "A History of the Development of the Boise Irrigation Project." Masters Thesis, Brigham Young University, 1969.

Chaffee, Eugene. "Early History of the Boise Region, 1811–1864." Masters Thesis, University of California, Berkeley, 1931.

Weatherby, James Benjamin. "The Hells Canyon Controversy: A Study of the Hells Canyon Associations and their View of Comprehensive River Basin Development." Masters Thesis, University of Idaho Graduate School, June 1968.

Gutema, Yoseph. "An Exposte Study of the Economic Performance of Federal Investments in Flood Control Projects in the Boise Valley, Idaho." Masters Thesis, University of Idaho, August, 1977.

Murphy, Paul L. "Irrigation in the Boise Valley, 1789–1869: A Study in Pre-Federal Irrigation." Masters Thesis, University of California, Berkeley, 1948.

Stine, Jeffrey. "Environmental Politics and Water Resources Development: The Case of the Army Corps of Engineers During the 1970s." Ph.D. dissertation, University of California, Santa Barbara, 1984.

## Newspapers

*Caldwell Tribune*
*Idaho Business Review*
*Idaho Daily Statesman*
*Idaho Free Press*
*Idaho Statesman*
*Idaho Tri-Weekly Statesman*
*Intermountain Observer*
*New York Times*
*Oregon Journal*
*Statewide*

## Archival Collections and Governmental Files

Ada County Office of Civil Defense
City of Boise
    Department of Planning and Community Development
    Park Department
    Public Works Department
    Office of the City Clerk
Morrison-Knudsen Records Center, Boise, Idaho
Seattle Federal Records Center
    National Archives and Records Administration
    Record Group 77
Corps of Engineers, Walla Walla District Office
Bureau of Reclamation, Region 1, Boise Idaho
State of Idaho
    Department of Lands
    Fish and Game Department
    Department of Water Resources
    Idaho State Historical Society
    Papers of Governor C. A. Bottolfsen
    Papers of Governor C. A. Robins
    Papers of Senator Henry Dworshak

## Interviews

William Ancell
Jack Blake
Ron Barrett
Gordon Bowen
Jack Cooper
Alice Dieter
Dick Eardley
Kathleen Gager
Ron Golus
Charles Hummel
Chris Korte
Charles Mickelson
Leo Ed Miller
Peter O'Neill
David Reese
Earl Reynolds
Robert Rickel
Alvin Ross
John Roten
Thomas Slater
Charles Winder

# Index

Ada County: 17, 26, 77, 80, 95; Board of Commissioners, 7, 17, 57, 79, 121; Civil defense, 107; development standards, 95, 100; Flood of 1983, 107, 113–114, 127; Highway District, 90
Ada County Fish and Game League, 57
Agriculture: 25–26, 36, 40, 71, 125; in flood plain, xxi, xxiii, 6, 8, 80; flood damage to, 5, 15, 29, 111
Aiken, 4
Albertsons, 84, 86
American Legion, 78
Americana Bridge, 78, 95
Ancell, William, 59–60, 62
Anderson, C.C., 26
Anderson Ranch Dam/Reservoir: 31, 52; flood control, 2, 10, 17, 30, 35, 111; Mountain Home Project, 27, 31, 37; and Boise River management system, 29, 31, 48, 64–65
Andrus, Cecil D., 58, 60
Ann Morrison Park, 68, 71–72, 73, 78, 95
*Annual Report, 1967* (Boise Park Department), 74
Army, U.S., 12, 26
Arrowrock Dam/Reservoir: 9, and Boise River management system, 29, 31, 33, 34–35, 48, 51, 65; construction, 2, 8, 10, 52; flood control, 11–12, 30, 47, 111–112, 119; irrigation, 25; and Lucky Peak pool, 1, 36, 60–61
Athens, 119
Atkinson, Harold E.: 72, 74; Report/plan, 74, 80
Auxiliary channel, 86, 89, 92, 105

Bald eagle, xxiii, 97–99, 103
Ballou, Ervin, 114

Barber Bridge, 79
Barber Park, 79, 107, 109
Barrett, Ron, 86, 87, 96, 113, 114, 124
Beattie, Charles F., 21, 22
Bench (area of Boise), xviii, 6, 8
Benefit/cost: bridges, 125; of "flushing flows," 120; and irrigation, 29; levees, 20, 29, 58; Lucky Peak, 16, 31–38, 123; Mountain Home Project, 27
Big Wood River, 107
Bike path: 76–79, 87–93, 98–104; 1983 flood, 108, 111, 116
Blake, Jack, 107, 115, 127
Boise Canyon, 15
Boise Chamber of Commerce, 40, 78
Boise (city): citizens, xxi, xxii, 15, 128; flood damage, 20, 27, 106, 108, 111–116, 121–122, 125; flood policy, 21–22, 74–77, 80–83, 94–103, 122; growth, 10, 67, 80, 85, 104, 125; land use, 2, 5, 6, 7, 11, 72–74, 119, 125; minimum flow/second outlet tunnel, 44, 48, 51–52, 54, 58–62, 65; National Flood Insurance Program, 80–81, 103, 106, 113, 121, 125; ParkCenter, 83–86; population, 10, 48, 58; river clean–up, 78–79; River Run, 87–94. *See also* Boise River water quality, Environmental impact statement, Flood plain, Floods of Boise River, Greenbelt
Boise City Council: 72, 74–76, 79; Crandlemire, 97; and NFIP, 81; and flood–plain policy, 83–85, 94, 98–99, 124; River Run, 91, 93
Boise Front, 10
Boise High School, 78
Boise Jaycees, 78
Boise Junior College, 71, 72
Boise National Forest, 107

Boise Park Department, 71, 74, 107–108
Boise Planning and Zoning Commission: 74; Crandlemire, 96; Flood plain policy, 83, 84, 93, 99, 124; Forest River, 95; River Run, 87
Boise Project: 8, 27; Board of Control, 10, 57, 60, 61
Boise Public Works Department, 59, 128
Boise River: aesthetics, xxi, 68–74, 103–104; auxiliary channel, 86, 87, 105, 110; pre-Lucky Peak flooding, xxi–xxii, 15; post-Lucky Peak flooding, 55–58; channel capacity, 10, 12, 17, 18, 47, 56–57, 70, 97, 100, 106, 108, 118, 119–120; channel clearing, 6, 18, 19, 30, 105, 115; channel straightening, 7, 10, 17, 19, 21, 121; conflict over water, 44; description, xxi, 2, 4–5; discharge/flow, 5, 11, 29, 33, 47, 56, 63–64, 106–108, 112, 121; diversions, 2, 4, 6, 8, 12, 18, 27, 47–48, 54, 86, 119; fishery, 51–55, 70, 92, 128; flood-plain development, 80–106; establishment of Greenbelt, 67–79; irrigation of Mountain Home desert, 27, 28, 31–32, 37; levees, xxii, 20, 22, 29, 55–58, 81–82, 86–105; location of flood plain, 75; Lucky Peak operations, xxi–xxiii, 1, 29, 31–38, 45–66; management responsibility, 121–122; map, 3; minimum flow, 51–65, 68–70, 124; and quality of life, 79; rafting, 68, 71, 72, 79; shutoffs, 51–55, 60, 68; South Fork, 10, 107; water quality, 11, 19, 44, 58–59, 67–71, 77; water rights, xxiii, 4, 6, 10, 51–55, 62. *See also* Bridges, Corps of Engineers, Flood plain, Flooding, Floods of Boise River, Greenbelt, Levees, Mores Creek
Boise River Plan, 96–100, 106, 120
Boise River Watch Committee, 96
Boise State College, 78
Boise State University, 76, 121
Boise Valley: xxiii, 5, 26, 35; agriculture, 8, 9, 11, 19, 21; description, xxii–xxiii; flooding, 1, 2, 15–17, 41; flooding after Lucky Peak, 55–58; irrigation interests, xxiii, 21–23, 25, 40, 44, 61, 64–66, 123, 128; population, 9, 10, 80. *See also* Floods of Boise River
Boise Water Company, 20
Bon Marche, 26
Bonneville Dam, 23
Bonneville Power Administration, 60
Bottolfsen, C.A., 12, 15, 21, 24
Bridges: 112, 121; and channel straightening, 10, 18; flood damage, 7, 8, 14–16, 20–21, 29, 30, 125; flood fight, 14; foot, 74, 76, 77, 90. *See also* names of bridges
Broadway Avenue, 15

Broadway Bridge, 6, 7, 14, 72, 83, 121
Brown, James, 1, 42
Brush, Harold, 106
Bureau of the Budget, 40
Bureau of Reclamation: 19, 59, 121, 128; construct dams, 1, 10; and Corps, xxii, 22–23, 32–38, 45–48, 62–63, 65; and fishery in Boise River, 51–55; and Flood of 1983, 106; Mountain Home Project, 26–27, 33, 35, 55
Burns, Stanley M., 57

Caldwell, 8
California, 72
Canyon County: 18, 58, 77, 108, 112; Board of Commissioners, 8, 17, 26, 57
Carscaddon, Joan, 108
Cascade Reservoir, 27
Central Addition, 7, 8
Channel Road, 108
Church, Frank, 60
Clark, D. Worth, 24, 29
Clarke, Frederick J., xxii
Clean Water Act, 77
Columbia Basin, 22, 38
Columbia River, xxi, 1, 23, 27, 37, 38
Columbia Valley Authority, 37, 38
Congress: appropriates funds for Lucky Peak, 1, 39; authorizes Lucky Peak, xxi, 1, 2, 35–39, 46, 66; authorizes second outlet tunnel, 60–61; Commerce Committee, 24, 29, 36, 38; and Corps, xxiii, 10, 15, 22–23, 33–34, 55; and cost of flood relief, xxii, 75, 80; House, 35; Mountain Home Project, 32; responsibility for flood control, 122–123; Senate, 23, 35
Cooper, Jack, 91, 115–116
Corps of Engineers: and Anderson Ranch Dam, 10; Board of Consultants, 45; Board of Engineers for Rivers and Harbors, 35; competition with Bureau, 22–23, 37–38; constraints on management, 45; constructs Lucky Peak, xxi, 1, 39–43; estimate discharge of early floods, 5, 12; estimate 1943 flood damage, 15–16; and fishery in Boise River, 51–55; and flood control policy, xxi–xxiii, 10, 22, 83, 98–99, 104–105, 124; Flood of 1983, 107–118; Flood Plain Information Services Programs, 75; flood predictions, 114; Hydrologic Engineering Center, 82; justification for Lucky Peak, 32–35, 67; and levees, 14, 17, 19, 20, 56–57, 81–82, 95; management of Lucky Peak, 44, 45–51, 55, 63–66, 67, 71, 97; memorandum of agreement, 47–48, 55–56, 62, 65; and

# Index

National Environmental Policy Act, 59–62, 124; non–structural flood control, xxii, 128; North Pacific Division, 22, 37–38, 59; and Plantation, 101; and post–Lucky Peak flooding, 55–58; Portland District, 30; proposal for Lucky Peak, 29; and River Run, 97–94; second outlet issue, 58–62; "spheres of influence," 38, 39; technical assistance, 77, 81–87, 89, 93–96, 101–105, 106; views after 1943 flood, 22, 24, 29; Walla Walla District, xxii, 51, 52, 56, 59, 65, 75, 87, 98–99, 108, 124. *See also* Benefit/cost, Floods of Boise River
Crandlemire, Roger, proposal, 96–97

Davis, Tom, 2
Davis, California, 82
DeCoursey, T.M., 26
Deer Flat Reservoir, 8, 11–12. *See also* Lake Lowell
*Definite Project Report on Lucky Peak Dam, Boise River, Idaho,* 36–37
Depression, 10, 26
Dieter, Alice, 74
Discovery State Park, 79
Diversion Dam, 8, 9, 27, 29, 47n, 52
Dobbs, J.M., 17
*Downstream Channel Requirements,* 56
Driscoll, Lynn, 26, 40
Drought, 10
Dworshak, Henry, 15, 32, 56

Eagle (city), 13, 79
Eagle Island, 79, 108, 111
Eagle Road, 13
Eardley, Dick, 60, 62, 80, 83, 91, 95, 104
Eighth Street, 11,
Eighth Street Bridge, 95
Eisenhower, Dwight D., 41.
Emkay, 83–86, 87, 93, 99–100, 110
Environmental impact statements, 59–62
Environmental Protection Agency, 58–59, 128
Erosion: 11, 19, 56, 96, 108; Flood of 1983, 115–116, 118

Fairview Avenue Bridge, 14–15, 16, 77
Federal Emergency Management Agency,(FEMA): Flood of 1983, 110, 113; flood–plain regulations, 83, 85, 93–94, 101; manager of NFIP, 81, 121
Federal Energy Regulatory Commission, 128
Federal Water Pollution Control Act Amendment of 1972, 77
Ferejohn, John, 123

First Security Bank, 26
Fish and Game Department. *See* Idaho Fish and Game Department
Fish and Wildlife Coordination Act of 1946, 55
Fish and Wildlife Service, U.S., 51, 98
Flood control: reservoirs, 20, 21, 31–32, 57, 107, 118; zoning, xxii, 74–76, 81; nonstructural, xxi–xxii, 124, 128
Flood Control Act of 1946, 36
Flood Control Act of 1950, 56
Flood Control Districts, 57
Flood Control Policy, Task Force on Federal, xxii
Flood control policy, 10, 20, 80–106. *See also* Boise city, Corps of Engineers
Flood damage: xxi–xvii, 19, 20, 23, 44, 125; early floods (*See* Floods of Boise River); post Lucky Peak, 55–58, 106; Corps predictions, 112, 114; flood of 1983, 111, 113, 114, 115; liability, 116, 121; cost of, 75, 81, 94
Flood fights, 4, 5, 12, 20, 56. *See also* Flood of 1983 and other floods
Flood forecasting, 12, 116–117
Flood management, 7–8, 116, 123. *See also* Floods of Boise River, Corps of Engineers
Flood, 100–year: xxi, xxii, 21, 108, 118, 120; Corps estimates, 101; FEMA regulations, 81–83; flood–plain information, 75; Loggers Creek, 86; post-Lucky Peak level, 117
Flood plain: 19, 20, 37; development, 67, 74–76, 80–106, 117–119, 121, 124; floodway/floodway fringe, 81–85, 89–90, 93–97, 99, 104, 122; Garden City, 100–103; ParkCenter, 83–86; River Run, 87–94, 117. *See also* Corps of Engineers, Floods of Boise River, Greenbelt, Recreation, Urban growth
*Flood Plain Information, Boise, Idaho, and Vicinity,* 75, 81, 87
Flood Plain Information Service Programs, 75
Flood–proofing, 21, 93, 94
Floods of Boise River: 38; of 1876, 1, 2–5, 8; of 1896, 1, 2, 6–8; of 1936, 10, 18; of 1938, 18; of 1943, 1, 2, 8–16, 29, 47, 75, 106, 113, 114, 116, 123; local responses to 1943 flood, 17–22, 25; federal responses to 1943 flood, 22–23, 101, 106–118, 120, 121
Floods of other rivers: Columbia River, xxi, 37; Ohio River, xxi, 1936, 10; Susquehanna, xxi; Trinity, xxi
Ford, Gerald, 60
Forest River, 95
404 permit, 96, 101, 121

Garden City: 79, 104, 116; development in

flood plain, 100-103; flood of 1983, 111, 113-114
Garden Valley 27, 39
Glenwood Bridge, 47, 101
Goff, Abe McGregor, 38
Golden Rule department store, 26
Gowen Air Field, 12,
Grand Coulee Dam, 23, 38
Greenbelt: xxi, 97, 106, 122, 124, 126; development of, 67-79, 80, 84-85; River Run, 87-94; Boise River Plan, 98-99; Flood of 1983, 101; and flood-plain policy, 83, 103-104; and State of Idaho, 74, 76-79, 100-03
Greenbelt Committee, 75, 88-89, 91, 96

Hathaway, John, 98
Health Department, U.S., 52
HEC-II step-backwater program, 82, 86, 101, 108
Hells Canyon (high) Dam: 23, 41; SWIWCP interest in, 32, 40, 44; Corps/Bureau conflict, 37-38
Heritage Conservation and Recreation Service, 91
Hillcrest Canal, 27
Holt, C. Leo, 57
Hoover Dam, 26,
Horseshoe Bend Canal, 27
Hotel Boise, 29, 30
Housing and Urban Development, U.S., 81
Howard, Potter, 40
Hummel, Charles, 68

Idaho (state of): and Boise River Greenbelt, 74, 76-79, 100-03; Congressional delegation, 24, 35, 38, 35-36, 38, 40; growth, 25, 27, 40; irrigation interests, 32, 35, 42, 60; leaders, 26; recreation contract at Lucky Peak, 71. *See also* names of state agencies
Idaho City, 2
Idaho state constitution, 53
*Idaho (Daily, Triweekly) Statesman*: 106-107, 113, 116; editorial criticism, 1, 2, 42; flood coverage, 2, 4, 7; Greenbelt, 93, 103
Idaho Department of Parks and Recreation, 79
Idaho Department of Public Lands, 78
Idaho Department of Water Resources, 58, 59, 60, Board, 114
Idaho Fish and Game Department: 62, 65, 128; and flood-plain developments, 92, 96-98; hatchery, 14; reports on water quality, 70; seeks minimum flow, 51-55, 57
Idaho Land Board, 102-103

Idaho Legislature, 10, 18, 74
Idaho Power Company, 26, 41
Idaho Department of Transportation, 10
Idaho State Highway 21, 36
Idaho State Reclamation Association, 36
International Dunes Motel, 78-79
Irrigation: canals, 19, 112; development in Boise Valley, 5-6; districts, 10; diversions, 2, 4, 8, 19, 47, 78, 85-86, 111, 119; and flood control, 44, 117, 121; interests, 21, 22-23, 31-32, 53-55, 55, 62-66, 128; Kings River Project, 122-123; legislation, 8; and Lucky Peak, 31-35, 38, 40, 48; Roanoke River Basin Project, 123; storage of water, 1, 6, 10, 17, 25, 46
Island, The, 90, 104. *See also* River Run

Johnson, Lyndon, xxii
Julia Davis Park, 76

Kelly, Milton, 4, 5
Kent, John, 19
Kings River Project, 122-123
K-Mart, 95
Korean War, 39, 40, 41
Korte, Chris, 95
Krogh, Lee
Kulp, Mark, 18, 53-54

Lake Heron, 88, 89, 90. *See also* River Run
Lake Lowell, 8, 11-12, 47n, 53. *See also* Deer Flat
Land and Water Conservation Fund, 74, 91
Land use planning, 2, 72, 74. *See also* Flood plain development, Boise River Plan, Zoning
Lander Street Sewage Treatment Plant, 52, 55, 69
Leehey, Donald J., 17
Leonard, Ross, 52
Lewis, Oliver, 15
*Levee Restudy*, 58
Levees (dikes): 11, 15, 108, 111; benefit/cost, 20, 29, 58; built in 1970s, xxi-xxii, 92, 86-105, 106, 119-120, 124; City of Boise policy, 94, 99, 104-105; and Corps, 10, 12, 14, 17, 19-22, 56-57, 81-83; Crandlemire, 96; early flood control, 4-8; federal regulations, 81-82; Flood of 1983, 111-112, 116, 118; Forest River, 95; ParkCenter, 86, 94, 104; River Run, 89-90, 104, 117. *See also* Benefit/cost, Boise River, Corps of Engineers
Linder Bridge, 13
Little Wood River, 107
Loggers Creek, 85-86, 87

Loggers Creek Park, 89
Lucky Peak Reservoir/Dam: xxi, xxiii, 19, 30, 105; benefit/cost, 27–29, 31–38, 123, 125; and Congress, 1, 2, 35–39; construction, 39–43; Corps/Bureau competition, 22–23, 37; Corps planning for, 29–32; dedication, 40–42; environmental impact statement, 59–62, 124; federal justification for, 32–35, 125; impact on Boise River, 67–70, 106, 119; installation of hydroelectric power, 51, 59–62, 65; irrigation water, 44; and Mountain Home Project, 40; name, 30; operating agreement, 45–51; operations, xxi–xxiii, 1, 29, 31–38, 45–66, 67; operations in Flood of 1983, 101, 107–112; opposition to, 1, 30–31, 41–42; outlet tunnel, 45–51, 46, 52, 58–62, 65; plan of dam, 50; public hearing in 1944, 2, 15, 29, 33; and recreation, 71, 79; reduces flood risk, 106, 117–118, 125; rooster tail, 42; rule curves, 48, 49, 56, 58, 63–64, 106, 117; survey report, 15, 33, 35; SWIWCP lobbies for, 24–29, 32, 40, 42; *See also* Arrowrock, Boise River, Boise Valley, Congress, Corps of Engineers; Flood, 110-year; Floods of Boise River, Flood damage, Irrigation
Lumber mills: 4, 5, 6, 8, 67; McCaslin, 20; Rossi, 7

Maass, Arthur, 122–123
Macco–Puget Sound, 39
McClure, James, 61, 62
Main Street (Boise), 2, 7
Main Street Bridge, 114
Marsden, Al, 102
Meridian, 1
Metro Plan, 94–95. *See also* Boise River Plan
Mickelson, Chuck, 95, 110, 120
Military garrison, 2
Miller, A.H., 52, 53
Mills, W.H., 51, 52
Mines and mining, 2, 11, 18, 19, 119
Minimum flow: 44, 48, 51–52, 54, 58–62, 65; one of Lucky Peak functions, 67, 124
Mockwitz, Margaret, 104, 114
Mores Creek: 19, 20, 29, 36; Bridge, 36, 41–42
Morris, William B., 6
Morrison, Harry W., 24–26, 32, 40, 41, 71
Morrison–Knudsen Company, 26, 39, 46, 78, 83. *See also* Emkay
Mother Nature, 45, 47
Mountain Home, 26, 27, 31
Mountain Home Desert, 24, 26–27, 31, 40, 55
Mountain Home Project: 28, 32, 37–38; Bureau of Reclamation, 26–27, 32–33, 35, 55, 64; Hells Canyon, 40–41
*Muddy Waters*, 122
Multiple-purpose: projects, xxi, 10, 22–23, 40–41; Corps/Bureau agreement, 38; and Lucky Peak, 29, 31–33, 44, 67, 123
Municipal Golf Course, 76, 110–111
Municipal Park, 71, 115–116
Myrtle Street Neighborhood, 7

National Environmental Policy Act of 1969, 59, 124
National Flood Insurance Program (NFIP): 106, 121, 124, 125; emergency phase, 80–81; flood–plain regulation, 84, 93, 101, 103; insurance cost, 94–95, 102
National Guard, 78
National Reclamation Association, 26
National Weather Service, 107, 111
Nelson, Arlo, 74
New York Canal, 8, 12, 47n, 108
*New York Times*, xxii
Newell, Robert J., 37
Newell–Weaver agreement, 38
Newlands Reclamation Act of 1902, 8
Ninth Street Bridge, 3, 5, 10
Nortner, S.E., 35
Notus Bridge, 21

Ohio River, 10
Old Soldiers Home, 14
O'Neal, Tip, xxiii
O'Neill, Peter, 89, 91
Oregon, 1, 32
Overton, John Holmes, 36

Pacific Coast, 29
Pacific Northwest, 23, 60
Pacific Ocean, 1, 47
ParkCenter: 87, 89, 94, 103; development of, 83–86, 92, 95; flood of 1983, 108, 110, 115; Park, 85
ParkCenter Boulevard, 110
Parks. *See* names of parks
Parma, 21
"Partnership concept," 41
Payette River, 27, 28, 35, 37, 107
Peterson, Tony, 87
Pilfering, 14
Plantation (Development), 100–102, 111
Plantation Golf Course, 11, 14
Population, 2, 9, 10, 58, 80
*Pork Barrel Politics*, 123
Portland, Oregon, 1, 37, 45

Post Office, U.S., 95
Power (hydroelectric): Corps 308 reports, 22–23; installation at Lucky Peak, 51, 59–62, 65; Mountain Home Project, 27, 31–40; multipurpose projects, 17, 44, 48, 123, 128
Public Health Department, U.S., 51
*Public Law 84–660*, 59

Quick, Paul, 51

Reclamation. *See* Irrigation
Reclamation Service, xxii, 2, 8, 10. *See also* Bureau of Reclamation
Recreation: 128; and Boise (city) flood-plain policy, 81, 83, 93–94, 98, 103, 124; in and along Boise River, 68, 71, 74–75, 79, 106; Lucky Peak benefit, 34, 37; Lucky Peak operations, 44, 46, 48, 64, 67, 121
Reed, Scott, xxiii
Reese, Dave, 65–66
*Reservoir Regulation Manual*, 47, 48, 58, 62–67, 128
Reuss, Martin, xxi
*Review of Survey Report*, 35, 36
Ridenbaugh Canal, 6,
Rising, Everett W., 26, 32, 38, 39, 40, 128
River Run: 95, 103, 104, 117; development, 87–94, 90. *See also* Island, The
Riverside Village, 102–103, 104, 108, 111, 112, 114
Roanoke River Basin Project, 123
Robins, C.A., 38, 40
Robins, Thomas M., 32
Roderick, George H., 40
Rooster tail, 42, 43
Ross, C. Ben, 11
Rossi lumber mill, 7
Roughness coefficient, 85

Salmon River, 27, 107
Salt Lake City, 107, 112
Sawtooth Mountains, 1
Seabees, 78
Secretary of the Army, 47
Secretary of the Interior, Assistant, 47
Selander, Glenn, 98
Shabman, Leonard, 123
Shiloh Inn, 78–79
Sierra Club, 78
Siltation: 29, 37, 51, 111, 119–120; mining as source, 18, 19
Simplot, J.R., and Company 26
Smead, J.R., 52
Smith, Charles, 111

Snake River: 28; basin system, 1, 5; Corps/Bureau agreement, 38; flooding, 12, 107; and irrigation, 26–27, 47; 308 reports, 23. *See also* Hells Canyon
Snake River Plain, 26
Snotel, 63, 64
Soil Conservation Service (U.S. Dept of Agriculture), 19, 63, 125
South Fork of Boise River, 10, 107
Southwest Idaho Water Conservation Project, Inc.(SWIWCP): 128; goals, 33, 40, 44; lobby for Lucky Peak, 24–33, 36, 38, 39, 42
"Spheres of influence," 39
Sporseen, Stanley, 29
Stanfield, Rochelle, 128
Star Bridge, 15
State Street, 11
*Statesman*. *See Idaho Daily Statesman*
*Statewide*, 41–42
Steele, Chuck, 113
Strauss, Michael, 36
Sweepy–Weepy. *See* Southwest Idaho Water Conservation Project, Inc.

Taubman Corporation, 74
Tennessee Valley Authority, 37
Terteling, 39
308 Reports, 22–23, 37
321st (Reserve) Engineer Battalion, 78
Thucydides, 119
Triangle K construction yard, 14
Trout fishery: 44, 70, 128; and flood plain development, 91, 92, 96, 98, 104
Truman, Harry S., 36, 37, 40
Tudor, Ralph, 30

Unterkoefler, Gerry, 87
Urban growth: xxiii, 25, 27, 42, 58, 125; east of Boise, 85; effect of irrigation development, 10; Garden City, 100; 1970s–1980s, 67, 80, 103–04. *See also* Population, Boise (city)
Utah, 107

*Valleys of Tomorrow*, 25
Vanport, 37
Veterans Park, 14
Volunteer Reserves, 14

Walker, Austin, 15, 20
Walla Walla District: xxii, 65, 108, 124; EIS, 59; flood information service, 75, 87, 98; and minimum flow, 51–52; post–Lucky Peak flooding, 56. *See also* Corps of Engineers
War Department, 24

## Index

War Production Board, 10, 17,
Washington, DC, 26, 37
Water Resources Development Act of 1974, xxii
Water Resources Development Act of 1976, 60
Watershed management, 19
Weaver, Theron D., 37
Webb, William, 70
Welsh, William: 12, 17, 21, 22; and SWIWCP, 26, 29, 36, 38
West, American, xxii–xxiii, 2, 10, 23, 27, 60, 128
West Boise Sewage Treatment Plant, 62, 63, 77
Western Beet Growers Association, 26
Westpark Shopping Center, 100

Wetherell, Robert, 26
White, Robert M., 18
Whitten, R.E., 30, 31, 41
Williams, Robert, 108
Winder, Charles, 110
Works Progress Administration, 10,
World War II, 10, 11, 22, 23, 26, 123
Worster, Donald, xxii–xxiii

Yost, Harry L., 26
Youtz, Jeff, 114

Zoning: general, 72, 74, 76; greenbelt, 75–76, 79; shopping mall, 100